T0140217

Advances in Intelligent Systems and Computing

Volume 1010

The series "Advances in Intelligent Systems and Computing" contains publications on theory, applications, and design methods of Intelligent Systems and Intelligent Computing. Virtually all disciplines such as engineering, natural sciences, computer and information science, ICT, economics, business, e-commerce, environment, healthcare, life science are covered. The list of topics spans all the areas of modern intelligent systems and computing such as: computational intelligence, soft computing including neural networks, fuzzy systems, evolutionary computing and the fusion of these paradigms, social intelligence, ambient intelligence, computational neuroscience, artificial life, virtual worlds and society, cognitive science and systems, Perception and Vision, DNA and immune based systems, self-organizing and adaptive systems, e-Learning and teaching, human-centered and human-centric computing, recommender systems, intelligent control, robotics and mechatronics including human-machine teaming, knowledge-based paradigms, learning paradigms, machine ethics, intelligent data analysis, knowledge management, intelligent agents, intelligent decision making and support, intelligent network security, trust management, interactive entertainment, Web intelligence and multimedia.

The publications within "Advances in Intelligent Systems and Computing" are primarily proceedings of important conferences, symposia and congresses. They cover significant recent developments in the field, both of a foundational and applicable character. An important characteristic feature of the series is the short publication time and world-wide distribution. This permits a rapid and broad dissemination of research results.

** Indexing: The books of this series are submitted to ISI Proceedings, EI-Compendex, DBLP, SCOPUS, Google Scholar and Springerlink **

More information about this series at http://www.springer.com/series/11156

Javier Prieto · Ashok Kumar Das ·
Stefano Ferretti · António Pinto ·
Juan Manuel Corchado
Editors

Blockchain and Applications

International Congress

 Springer

Editors
Javier Prieto
BISITE Research Group
Salamanca, Salamanca, Spain

Stefano Ferretti
Dipartimento di Informatica - Scienza e
Ingegneria
University of Bologna
Bologna, Italy

Juan Manuel Corchado
BISITE Digital Innovation Hub
University of Salamanca,
AIR Institute - Deep Tech Lab
Salamanca, Salamanca, Spain

Ashok Kumar Das
International Institute of Information
Technology
Gachibowli, Hyderabad, India

António Pinto
Politecnico do Porto and INESC TEC
Porto, Portugal

ISSN 2194-5357   ISSN 2194-5365   (electronic)
Advances in Intelligent Systems and Computing
ISBN 978-3-030-23812-4   ISBN 978-3-030-23813-1   (eBook)
https://doi.org/10.1007/978-3-030-23813-1

This Springer imprint is published by the registered company Springer Nature Switzerland AG
The registered company address is: Gewerbestrasse 11, 6330 Cham, Switzerland

Preface

The 1st International Congress on Blockchain and Applications 2019 (BLOCKCHAIN'19), held in the Heritage city of Ávila, has been a forum for experienced and young researchers on blockchain and artificial intelligence (AI) where they have shared ideas, projects, lectures, and advances associated with these technologies and their application domains. Among the scientific community, blockchain and AI are seen as a promising combination that will transform the production and manufacturing industry, media, finance, insurance, e-government, etc. Nevertheless, there is no consensus with schemes or best practices that would specify how blockchain and AI should be used together. Combining blockchain mechanisms and artificial intelligence is still a particularly challenging task, and the BLOCKCHAIN'19 congress has been a milestone toward its achievement.

The BLOCKCHAIN'19 congress has been devoted to promoting the investigation of cutting-edge blockchain technology, exploring the latest ideas, innovations, guidelines, theories, models, technologies, applications, and tools of blockchain and AI for the industry, and identifying critical issues and challenges that researchers and practitioners must deal with in future research. The technical program has been carefully designed to offer a fresh and balanced selection of advances and results in blockchain and AI, encouraging the presence of fresh and interdisciplinary topics.

The technical program will present both high quality and diversity, with contributions in well-established and evolving areas of research. More than 40 papers were submitted to main and special sessions tracks from over 19 different countries (Canada, France, Germany, India, Ireland, Italy, Jordan, Luxembourg, Malaysia, Malta, Morocco, Netherlands, Oman, Portugal, Slovenia, Spain, Sweden, United Arab Emirates, and USA).

This symposium is organized by the University of Salamanca, IIIT Hyderabad, University of Bologna, and António Pinto - Instituto Politécnico do Porto (Portugal). This first edition will be held in Ávila, Spain, from June 26 to 28, 2019.

We thank the sponsors (IEEE Systems Man and Cybernetics Society Spain Section Chapter and the IEEE Spain Section (Technical Co-Sponsor), IBM, Indra, Viewnext, Global exchange, AEPIA, APPIA and AIR institute) and the funding supporting of the with the project *"Intelligent and sustainable mobility supported by multi-agent systems and edge computing"* (Id. RTI2018-095390-B-C32), and finally, the Local Organization members and the Program Committee members for their hard work, which was essential for the success of BLOCKCHAIN'19.

Javier Prieto
Ashok Kumar Das
Stefano Ferretti
António Pinto
Juan Manuel Corchado

Organization

General Chairs

Juan Manuel Corchado
    Rodríguez
University of Salamanca, Spain,
    and AIR Institute, Spain

Javier Prieto Tejedor
University of Salamanca, Spain,
    and AIR Institute, Spain

Program Committee Chairs

Ashok Kumar Das
IIIT Hyderabad, India

Stefano Ferretti
University of Bologna, Italy

António Pinto
Instituto Politécnico do Porto, Portugal

Program Committee

Imtiaz Ahmad Akhtar
IT Consultant, Sweden

Sami Albouq
Islamic University of Madinah, Saudi Arabia

Luís Antunes
Universidade do Porto, Portugal

Massimo Bartoletti
Dipartimento di Matematica e Informatica,
    Universita' degli Studi di Cagliari, Italy

Francisco Luis
    Benítez Martínez
University of Granada, Spain

Nirupama Bulusu
Portland State University, EE. UU.

Roberto Casado
University of Salamanca, Spain

Arnaud Castelltort
Montpellier, France

Liang Cheng
Lehigh University, USA

Giovanni Ciatto
University of Bologna, Italy

Denisa-Andreea
    Constantinescu
University of Málaga, Spain

Manuel E. Correia
CRACS/INESC TEC; DCC/FCUP, Portugal

Fernando De La Prieta
University of Salamanca, Spain

Josep Lluis De La Rosa	EASY Innovation Center, UdG & RPI, Spain
Roberto Di Pietro	Hamad Bin Khalifa University - College of Science and Engineering, Qatar
Ali Dorri	Qaen, Iran
Joshua Ellul	University of Malta, Malta
Miguel Frade	Instituto Politécnico de Leiria, Portugal
Hélder Gomes	Escola Superior de Tecnologia e Gestão de Águeda, Universidade de Aveiro, Portugal
Ramesh H. L.	Vidyavardhaka College of Engineering, India
Abdelhakim Hafid	University of Montreal, Canada
Marc Jansen	University of Applied Sciences Ruhr West, Germany
Raja Jurdak	Commonwealth Scientific Industrial and Research Organization, Australia
Aida Kamisalic	University of Maribor, Faculty of Electrical Engineering and Computer Science, Slovenia
Salil Kanhere	The University of New South Wales, Australia
Denisa Kera	Asia Research Institute (STS Cluster), Australia
Nida Khan	University of Luxembourg, Luxembourg
Mohamed Laarabi	Mohammadia School of Engineering Rabat, Morocco
Anne Laurent	LIRMM - UM, France
Jose Maria Luna	Dept. of Computer Science and Numerical Analysis, Spain
Fengji Luo	The University of Sydney, Australia
João Paulo Magalhaes	ESTGF, Porto Polytechnic Institute, Portugal
Qutaibah Malluhi	Qatar University, Qatar
Stefano Mariani	Università degli Studi di Modena e Reggio Emilia, Italy
Luis Carlos Martínez	University of Salamanca, Spain
Rolando Martins	University of Porto, Portugal
Imran Memon	Zhejiang University, China
Jelena Misic	Ryerson University, Canada
Anang Hudaya Muhamad Amin	Higher Colleges of Technology, United Arab Emirates
Daniel Jesus Munoz Guerra	University of Malaga, Spain
Agoulmine Nazim	usthb, Algeria
Andrea Omicini	Alma Mater Studiorum–Università di Bologna, Italy
Pedro Pinto	Instituto Politécnico de Viana do Castelo, Portugal
Matthias Pohl	Otto-von-Guericke-Universität Magdeburg, Germany
Yuansong Qiao	Athlone Institute of Technology, Ireland
Rogério Reis	University of Porto, Portugal

Esteban Romero-Frías	University of Granada, Spain
David Rosado	University of Castilla-La Mancha, Spain
Dinesh Saini	Sohar University, Oman
David Saive	University of Oldenburg, Germany
Altino Sampaio	Instituto Politécnico do Porto, Escola Superior de Tecnologia e Gestão de Felgueiras, Portugal
Ricardo Santos	ESTG/IPP, Portugal
Khaled Shuaib	College of Information Technology, UAEU, United Arab Emirates
Biplab Sikdar	National University of Singapore, Australia
Helder Sousa	Politécnico do Porto - Escola Superior de Tecnologia e Gestão, Portugal
Radu State	University of Luxembourg, Luxembourg
Subhasis Thakur	National University of Ireland, Galway, Ireland
Vicente Traver	Universitat Politècnica de València, Spain
Odelu Vanga	Birla Institute of Technology & Science (BITS), Pilani, Hyderabad Campus, India
Sebastián Ventura	University of Cordoba. Dept. of Computer Science and Numerical Analysis, Spain
Marco Vitale	Foodchain Spa, Italy
Stefan Wunderlich	University of Oldenburg, Germany
Zibin Zheng	Sun Yat-sen University, China
Kashif Zia	Sohar University, Sohar, Oman
Roberto Zunino	University of Trento, Italy
André Zúquete	University of Aveiro, Portugal

Organizing Committee

Juan Manuel Corchado Rodríguez	University of Salamanca, Spain, and AIR Institute, Spain
Javier Prieto Tejedor	University of Salamanca, Spain, and AIR institute, Spain
Roberto Casado Vara	University of Salamanca, Spain
Sara Rodríguez González	University of Salamanca, Spain
Fernando De la Prieta	University of Salamanca, Spain
Sonsoles Pérez Gómez	University of Salamanca, Spain
Benjamín Arias Pérez	University of Salamanca, Spain
Pablo Chamoso Santos	University of Salamanca, Spain
Amin Shokri Gazafroudi	University of Salamanca, Spain
Alfonso González Briones	University of Salamanca, Spain, and AIR Institute, Spain
José Antonio Castellanos	University of Salamanca, Spain
Yeray Mezquita Martín	University of Salamanca, Spain
Enrique Goyenechea	University of Salamanca, Spain

Javier J. Martín Limorti   University of Salamanca, Spain
Alberto Rivas Camacho   University of Salamanca, Spain
Ines Sitton Candanedo   University of Salamanca, Spain
Daniel López Sánchez   University of Salamanca, Spain
Elena Hernández Nieves   University of Salamanca, Spain
Beatriz Bellido   University of Salamanca, Spain
María Alonso   University of Salamanca, Spain
Diego Valdeolmillos   University of Salamanca, Spain,
                                   and AIR Institute, Spain
Sergio Marquez   University of Salamanca, Spain
Guillermo Hernández   University of Salamanca, Spain
  González
Mehmet Ozturk   University of Salamanca, Spain
Luis Carlos Martínez de   University of Salamanca, Spain,
  Iturrate   and AIR Institute, Spain
Ricardo S. Alonso Rincón   University of Salamanca, Spain
Javier Parra   University of Salamanca, Spain
Niloufar Shoeibi   University of Salamanca, Spain
Zakieh Alizadeh-Sani   University of Salamanca, Spain
Jesús Ángel Román Gallego   University of Salamanca, Spain
Angélica González Arrieta   University of Salamanca, Spain
José Rafael García-Bermejo   University of Salamanca, Spain
  Giner

Contents

IMBUA: Identity Management on Blockchain for Biometrics-Based User Authentication

Vanga Odelu[✉]

Birla Institute of Technology and Science Pilani,
Hyderabad Campus, Hyderabad 500078, India
odelu.vanga@hyderabad.bits-pilani.ac.in
https://sites.google.com/site/odeluvanga/home

Abstract. In centralized systems, privacy of user identity information relies on centralized entity, such as system administrator. Since centralized entity is responsible to distribute keys to all the users in the system, he can read and modify the identity information. Compromise of such centralized entity results into the entire system vulnerability. Therefore, centralized systems often failed to provide the data privacy and integrity. In addition, centralized systems are also vulnerable to single-point-of-failure. Blockchain based solution is a possible alternate to provide integrity and protect system from single-point-of-failure. But, the ledger of the blockchain, contains all the transactions information, is distributed and available to each member of the blockchain network. Hence, blockchain-based identity management for user authentication, while preserving privacy, is a challenging problem. This paper proposes a novel key management mechanism for blockchain-based identity management for user authentication. The rigorous analysis is presented to show that the proposed protocol is secure against various possible attacks.

Keywords: Blockchain · Identity management · Authentication · Biometrics · Security · Privacy

1 Introduction

Rapid development of Internet technology brings many security threats to the data privacy and integrity. In majority of the organizations, like universities and banks, the data privacy relies on a centralized single trusted entity, such as system administrator, who is responsible for key distribution to all the users in the system. Making changes to the identity information is easy because the whole system is under the single central-man control and non-irreversible. The data stored in centralized systems may reveal to adversary in a single point of compromise. Thus, centralized systems often failed to provide data privacy and integrity. In addition, centralized systems become a perfect target for cyber-attacks: a recent incident like Equifax breach[1], has affected more than 145 million

---

[1] http://fortune.com/2017/10/20/equifax-breach-credit/.

© Springer Nature Switzerland AG 2020
J. Prieto et al. (Eds.): BLOCKCHAIN 2019, AISC 1010, pp. 1–10, 2020.
https://doi.org/10.1007/978-3-030-23813-1_1

US citizens credit card information. Moreover, centralized systems are also vulnerable to the single-point-of-failure. Blockchain technology has the potential to provide integrity of the identity information without relying on the centralized trusted authority, and also secure against single-point-of-failure [1].

In 2008, Satoshi Nakamoto was first proposed a concept of Bitcoin as a cryptocurrency, and later in 2009, Blcokchain was implemented [2]. Blockchain is an immutable ledger implemented in a distributed fashion without a central authority (i.e., a bank, company or government) [1]. Each member of the Blockchain network is a computer node, which runs a specific client that allows to connect to the blockchain. When a node joins the Blockchain network, it is permitted to hold a full copy of the blockchain ledger. The process of validating the transactions by the network nodes is known as *mining* and those nodes are called *minors*. The minors mine the transactions and generate the blocks with valid set of transactions by reaching the consensus, using the consensus protocols, such as *proof-of-work, proof-of-stake, delegated proof-of-stake,* and *proof-of-importance,* etc., [3,4]. After the Bitcoin was introduced blockchain to the world, various types of blockchains has been emerged based on the permission models, such as Public permissionless blockchain (open to anyone): Bitcoin [2] and Ethereum [5]; Public permissioned blockchain (open to anyone and authorised participants can write): Sovrin [6]; Permissioned blockchain (restricted to authorised set of participants): Quorum [7] and Hyperledger [8]; Private permissioned blockchain (fully private or restricted to a limited set of authorised nodes): Bankchain [9].

Blockchain ledger is immutable, and therefore, provides integrity. However, ledger is distributed and public to all the legal network members. Therefore, identity management on the blockchain for user authentication, while preserving privacy, becomes an interesting and emerging research problem. This paper presents a novel key management mechanism to manage identity on the blockchain for user authentication. This is, to the best of our knowledge, the first privacy preserving identity management on blockchain for user authentication suitable for consortium blockchain.

Rest of the paper is organized as follows: System model is discussed in Sect. 2. In Sect. 3, discuss the required mathematical preliminaries, and propose a novel blockchain-based identity management and authentication protocol in Sect. 4. In Sect. 5, the rigorous security analysis of the proposed protocol is presented, and analyzes the performance in Sect. 6. Finally, paper is concluded in Sect. 7.

2   System Model

In this paper, the permissioned-blockchain network is considered. In permissioned blockchain, each legal member of the network (forum members) holds the distributed ledger. The proposed system consists of three major roles, namely, User ($\mathcal{U}$), Registration Center ($\mathcal{RC}$), Authentication Server ($\mathcal{AS}$). Initially, this system establishes a permissioned blockchain network with the trusted forum of members, includes the roles $\mathcal{RC}$ and $\mathcal{AS}$. A new member is allowed to join the network with the acceptance of majority of the existing forum members (consensus). $\mathcal{RC}$ is responsible to verify the user's identity information physically and push it into the blockchain as the transaction of *user-registration-request* for mining. After successful mining, user identity-information (*IDInfo*) will be updated onto the blockchain ledger. After updating *IDInfo*, $\mathcal{RC}$ issues authentication credentials to the user $\mathcal{U}$. The user uses these credentials to prove himself at any authentication server $\mathcal{AS}$ without the help of $\mathcal{RC}$. The summary diagram of the proposed system model and message flow is shown in Fig. 1.

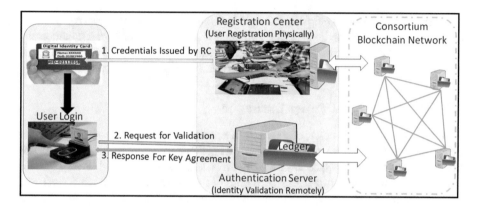

Fig. 1. System network model

3 Mathematical Preliminaries

The required mathematical preliminaries are discussed in this section.

3.1 Bilinear Pairing

Let two groups G and G_T are additive and multiplicative cyclic groups, respectively, with the same prime order p. Suppose P is a generator of G and map $e : G \times G \rightarrow G_T$, called a bilinear pairing, satisfies the following properties [10].

- $e(aP, bP) = e(P, P)^{ab}$, $\forall P \in G$ and $\forall a, b \in Z_p^*$.
- $e(P, P) \neq 1$, where $1 \in G_T$ is the identity element.
- Computing the pairing e is efficient.

We denote the pairing group as $\{e, G, G_T, P, p\}$.

Definition 1 (Bilinear Diffie-Hellman Problem). *Given P, aP, bP, and cP in G, for some $a, b, c \in Z_p^*$, computing $e(P, P)^{abc} \in G_T$ is infeasible. Equivalently, for given P, aP, bP, and cQ for some $a, b, c \in Z_p^*$, it is infeasible to compute $e(P, Q)^{abc} \in G_T$, where Q generated with secure map-to-point hash function $H : \{0, 1\}^* \rightarrow G$, that is, $Q = H(IDInfo)$, where $IDInfo$ denotes Identity Information.*

Definition 2 (Computational Diffie-Hellman problem [11]). *Let g in G_T be a generator. Given random input g, g^a, g^b in G_T for some $a, b \in Z_p^*$, computing g^{ab} is a computationally infeasible problem.*

Definition 3 (Divisible Computation Diffie-Hellman problem [11]). *Let g in G_T be a generator. Given random input g, g^a, g^b in G_T for some $a, b \in Z_p^*$, computing $g^{b/a}$ is a computationally infeasible problem.*

Remark 1. The above Definitions 2 and 3 are equivalent in complexity over the cyclic group of prime order p [11]. Suppose g is a generator of G_T, a cyclic group of order p. From the above definitions, we can see the equivalent result that, given g^a, g^b, and g^c in G_T, computing $g^{bc/a}$ is a computationally infeasible.

3.2   Biometrics and Fuzzy Extractor

A fuzzy extractor $(\Upsilon, m, l, t, \epsilon)$ extracts a nearly l-bit random string σ from its input biometrics ω in an error-tolerant way, where m is the min-entropy of any distribution on metric space Υ and t the error tolerance threshold. Fuzzy extractor consists two procedures: (1) probabilistic generation procedure (Gen) and (2) deterministic reproduction procedure (Rep) [12]. Gen and Rep are defined as follows:

– Gen: For the input ω, gives the output $\langle \sigma, \theta \rangle \leftarrow Gen(\omega)$.
– Rep: For all $\omega, \omega' \in \Upsilon$ satisfying $dis(\omega, \omega') \leq t$, it produce $\sigma \leftarrow Rep(\omega', \theta)$.

User bimetric is unique and cannot be lost or forgotten. In addition, biometric keys are hard to forge or distribute, and also it is computationally infeasible to guess the biometric keys [13, 14].

4   Identity Management and Authentication Protocol

A novel blockchain-based identity validation and key agreement protocol is presented in this section. The detailed steps are as follows.

4.1   Initialization Phase

Initially, a consortium will be formed with the required roles, namely Registration Center ($\mathcal{RC}$) and Authentication Server ($\mathcal{AS}$). Each node in the consortium chooses bilinear pairing group $\{e, G, G_T, P, p\}$, a cryptographic hash function $h : \{0, 1\}^* \rightarrow Z_p^*$, and secure map-to-point hash function $H : \{0, 1\}^* \rightarrow G$.

– Role $\mathcal{RC}$ generates its private/public key pair, say $(r, R = rP)$ and joins the blockchain network using its public-key R.
– Role $\mathcal{AS}$ generates its private/public key pair, say $(a, A = aP)$ and joins the blockchain network using its public-key A.

4.2   User Registration Phase

In this phase, user $\mathcal{U}$ registers with the registration center $\mathcal{RC}$ using his/her personal biometrics. In the registration process, $\mathcal{RC}$ verifies the user identity physically, and then records the user's digital identity information on the blockchain. The detailed process is as follows:

R1. $\mathcal{U}$ inputs his personal biometrics Bio into the reader, fuzzy extractor (Sect. 3.2) enabled, to extract the biometric parameters $\langle \sigma, \theta \rangle$, that is, $\langle \sigma, \theta \rangle \leftarrow Gen(Bio)$. Note that σ is a secret parameter and θ is a public parameter. Next, $\mathcal{U}$ computes $U = uP$, where $u = h(\sigma, ts)$ and ts is current timestamp. $\mathcal{U}$ finally sends a registration request $\{U, IDInfo\}$ to the registration center $\mathcal{RC}$, where $IDInfo$ means $Identity\ Information$ which will be presented physically to $\mathcal{RC}$.

R2. After successful verification of the user $\mathcal{U}$'s $IDInfo$ documents (physically), $\mathcal{RC}$ assigns the user unique identity UID. $\mathcal{RC}$ then choose $t \in Z_p^*$ and computes $s = rh_t + t \pmod{p} \neq 0$, where $h_t = h(UID, T, R, U, V, lifetime, ts)$, $V = tU$, $Q = H(UID, R, V, U, lifetime)$, $T = tQ$. Next, $\mathcal{RC}$ sends user-registration-transaction $URT = \{UID, R, V, U, T, lifetime, ts\}$ to the blockchain mining. Upon successful mining, URT will be updated in the blockchain ledger. Finally, $\mathcal{RC}$ will issue a smart card $SC = \{UID, s, P, T, Q, lifetime\}$ to the user $\mathcal{U}$.

R3. After receiving SC from $\mathcal{RC}$, user $\mathcal{U}$ stores θ into SC, that is, SC finally consists $\{\theta, UID, s, P, T, Q, lifetime\}$. Note that s will be protected with the biometric secret σ.

Remark 2. In Step R2, $\mathcal{RC}$ composes a digital identity information URT and sends into the blockchain network for mining. Since each network member has the complete blockchain ledger, it is not hard to verify uniqueness of digital identity information URT while mining. Therefore, in the proposed network model, use of centralized identity generator is not required. One possible way of generate unique identity UID is using hash as $UID = h(r, R, IDInfo, ts)$, where (r, R) is private and public key pair of $\mathcal{RC}$, $IDInfo$ is identity information physically presented to $\mathcal{RC}$, and ts is current timestamp. In this case, pair (r, R) is unique to $\mathcal{RC}$, and remaining information in the hash may not match exactly with the other user's profile. Therefore, probability of matching the generated UID with the other UID' is negligible because of the collision-free property of hash function.

4.3   Identity Validation and Key Agreement Phase

In this phase, authentication server $\mathcal{AS}$ validates the user $\mathcal{U}$ based-on the identity information recored on the blockchain ledger. In addition, $\mathcal{AS}$ also establish a session key with the user $\mathcal{U}$. The detailed steps involved in this phase are discussed below.

A1. $\mathcal{U}$ inputs biometrics Bio' along with the smart card SC into the reader to compute the σ, that is, $\sigma = Rep(Bio', \theta)$. Next, user computes

$$x \leftarrow Z_p^*, \ X = xQ, \ k_1 = e(A, T)^u, Auth_1 = h(UID, X, k_1, ts_1, A)$$

Finally, $\mathcal{U}$ sends a verification request $\{UID, X, ts_1, Auth_1\}$ to $\mathcal{AS}$.

A2. After receiving request, $\mathcal{AS}$ retrieves the identity information $URT = \{UID, R, V, U, T, lifetime, ts\}$ corresponding to UID from the blockchain ledger and then checks the lifetime. If $lifetime$ of the URT is valid, then $\mathcal{AS}$ computes

$$k_2 = e(V, Q)^a, \quad \text{where} \ \ Q = H(UID, R, V, U, lifetime)$$
$$\text{Checks the validity of} \ Auth_1 =^? h(UID, X, k_2, ts_1, A).$$

If the above equation is valid, $\mathcal{AS}$ confirms that the identity of the user $\mathcal{U}$ is legitimate. Then $\mathcal{AS}$ proceeds the computations as follows for key agreement:

$$y \leftarrow Z_p^*, \ z = e(h_t R, Q)e(P, T) = e(P, Q)^{(h_t r + t)} = e(P, Q)^s, Y = z^y = e(P, Q)^{sy}$$
$$ek_2 = e(U, X)^y = e(P, Q)^{uxy}, Auth_2 = h(UID, X, U, Y, A, ts_1, ts_2, k_2, ek_2)$$

where $h_t = h(UID, T, R, U, V, lifetime, ts)$. Finally, $\mathcal{AS}$ sends $\{Y, Auth_2, ts_2\}$ as response to the user $\mathcal{U}$.

A3. Upon receiving response from $\mathcal{AS}$, the user $\mathcal{U}$ computes

$$ek_1 = \left(Y^{(u/s)}\right)^x = e(P, Q)^{uxy}$$
$$\text{Checks the validity of} \ Auth_2 =^? h(UID, X, U, Y, A, ts_1, ts_2, k_1, ek_1).$$

If the above equation is valid, $\mathcal{U}$ confirms the authenticity of $\mathcal{AS}$ and computes the following confirmation message. Otherwise, reject the session.

$$Conf = h(UID, k_1, ek_1, X, Y, ts_1, ts_2, Q, U, A).$$

Finally, $\mathcal{U}$ sends confirmation message $\{Conf\}$ to $\mathcal{AS}$.

A4. After receiving $\{Conf\}$ from $\mathcal{U}$, The authentication server $\mathcal{AS}$ checks the validity of the following equation:

$$Conf =^? h(UID, k_2, ek_2, X, Y, ts_1, ts_2, Q, U, A).$$

If the above equation is not valid, $\mathcal{AS}$ reject the session. Otherwise, $\mathcal{AS}$ confirms that $\mathcal{U}$ agree-on the common session key sk, where

$$sk = h(k_1, ek_1, ts_1, ts_2, UID) = h(k_2, ek_2, ts_1, ts_2, UID).$$

5   Security Analysis

In this section, rigorous security analysis is presented to show the proposed *IMBUA* protocol is secure against various attacks, and also supports the essential security properties such as credential reissue/update and non-repudiation.

Theorem 1. *Proposed protocol is secure against replay attack.*

Proof. $\mathcal{U}$ generates a validation request in Step $A1$ (Sect. 4.3) using the Bilinear Diffie-Hellman (BDH) Key $k_1 = e(A,T)^u = e(P,Q)^{tau} = k_2$, which is shared with $\mathcal{AS}$. Generating this key k_1 by any adversary $\mathcal{A}$ (including $\mathcal{RC}$), except $\mathcal{U}$ and $\mathcal{AS}$, is as hard as the BDH problem (Definition 1). In addition, computing confirmation message in Step $A3$ (Sect. 4.3) requires the long-term secrets u (generated with biometrics) and s (generated using ElGamal Digital Signature [15] by $\mathcal{RC}$). Therefore, without proper secrets, it is hard to success in replay attack. As a result, proposed protocol prevents the replay attack.

Theorem 2. *Proposed protocol is secure against man-in-the-middle attack.*

Proof. From Theorem 1, generating request (Step $A1$) and confirmation (Step $A3$) messages is hard to the adversary $\mathcal{A}$ without secret credentials u and s. Thus, the proposed $IMBUA$ also resists the man-in-the-middle attack.

Theorem 3. *Proposed protocol is secure against impersonation attack.*

Proof. In this attack, consider following two cases:

(1) $\mathcal{RC}$ **act as user** $\mathcal{U}$: In this case, $\mathcal{RC}$ needs to generate a valid request using BDH key $k_1 = e(A,T)^u = e(V,Q)^a = k_2$, which is shared between $\mathcal{U}$ and $\mathcal{AS}$. However, generating such a key by $\mathcal{RC}$ is equivalent to compute $e(P,Q)^{au}$ for given P, uP, aP (assume that t even stored at $\mathcal{RC}$ in worst case). In addition, confirmation message is computed as $Conf = h(UID, k_1, ek_1, X, Y, ts_1, ts_2, Q, U, A)$, where $ek_1 = \left(Y^{(u/s)}\right)^x = e(P,Q)^{uxy}$. In this case, adversary $\mathcal{A}$ (including $\mathcal{RC}$) have the possible tuple $\{e(P,Q)^s, z = e(P,Q)^{sy}, \text{ and } e(U,X) = e(P,Q)^{ux}\}$. However, computing key $ek_1 = e(P,Q)^{uxy}$ using the above available tuple is as hard as DCDH in G_T (Definition 3, Remark 1).

(2) $\mathcal{AS}$ **act as user** $\mathcal{U}$: As discussed in the above case, generating valid request and confirmation without the user secrets is infeasible to adversary includes $\mathcal{AS}$. Hence, the proposed protocol is secure against the impersonation attack by any adversary including $\mathcal{AS}$ and $\mathcal{RC}$.

Theorem 4. *Proposed protocol provides user privacy.*

Proof. In this case, proposed $IMBUA$ uses randomly generated unique identity UID for each registered user and store $URT = \{UID, R, V, U, T, lifetime, ts\}$ as identity information on the Blockchain, which is same as storing public-key in public blockchian. Therefore, proposed $IMBUA$ preserve the privacy of user on the blockchain.

Theorem 5. *Proposed protocol supports for user credential update.*

Proof. $\mathcal{U}$ generates secret key $u = h(\sigma, ts)$ using personal biometrics (fuzzy extractors), that is, $Gen(Bio) = (\sigma, \theta)$, where θ stored on the user card as public parameter to re-generate σ. Since, secret biometrics are not stored anywhere on the user card or blockchain directly, it will be protected from any adversary. In addition, if user card is lost, based-on the user $IDInfo$, the corresponding $\mathcal{RC}$ can find the user and reissue the new credentials using the same procedure because fuzzy extractors support for the cancelable biometrics [14].

Theorem 6. *Proposed protocol provides non-repudiation property.*

Proof. IMBUA protocol is designed based-on user personal biometrics. User biometric is unique, not forgeable, and difficult to copy [13]. In addition, *IMBUA* supports for revocation and reissue/update the user credentials when user unexpectedly lost the credentials, which will also record on the immutable blockchain ledger. Therefore, the proposed protocol ensures the non-repudiation property, that is, once a transaction complete and recorded successfully, cannot be denied.

6   Performance Analysis

In this section, we analyse the performance of the proposed *IMBUA* protocol in terms of computation cost for identity validation and key agreement at user and the authentication server. Suppose T_e, T_{Exp}, T_H, and T_h respectively denotes the execution timings of pairing, exponentiation in G and G_T, a map-to-point, and one-way hash function operations. We consider the following execution timings for various cryptographic operation presented in [16]. We assume that user $\mathcal{U}$'s device is HiPerSmart Card and Server $\mathcal{A}$ is Pentium IV.

Entity	T_e	T_{Exp}	T_H	T_h
Pentium IV	0.00316 s	0.00117 s	< 0.001 s	0.00001 s
HiPerSmart Card	0.38 s	0.13 s	< 0.1 s	< 0.1 s

Note: s denotes seconds

In *IMBUA*, the required computational overhead for $\mathcal{U}$ is $1T_e + 3T_{Exp} + 4T_h \approx 1.17$ s and $\mathcal{AS}$ is $4T_e + 2T_{Exp} + 1T_H + 4T_h \approx 0.01599$ s for the identity validation and key agreement phase. We assume that the size of UID, $h()$, and p is 160 bits and timestamp is 32 bits. Then, the communication overhead required in our protocol, for user $\mathcal{U}$ is $sizeof(UID, X, ts_1, Auth_1, Conf) = 160 + 320 + 32 + 160 + 160 = 832$ *bits*.

7   Conclusion

In this paper, a novel key distribution mechanism is proposed for blockchain-based identity management for user authentication, using personal biometrics. To the best of our knowledge, this is the first such protocol, which provides privacy to the identity information recorded on the blockchain ledger. Through the rigorous security analysis, it is proved that the proposed protocol is secure against various attacks, including replay, impersonation, and man-in-the-middle attacks. In addition, proposed protocol supports for user credential reissue/update when unexpectedly reveal/lost the credentials, and provides the non-repudiation property because biometric is unique. The computation overheads required for identity validation and key agreement are also analyzed in terms of execution timings of various cryptographic operations.

References

1. Yaga, D., Mell, P., Roby, N., Scarfone, K.: Blockchain technology overview. Technical report, National Institute of Standards and Technology (2018)
2. Nakamoto, S.: Bitcoin: a peer-to-peer electronic cash system (2008)
3. De Angelis, S., Aniello, L., Baldoni, R., Lombardi, F., Margheri, A., Sassone, V.: PBFT vs proof-of-authority: applying the cap theorem to permissioned blockchain (2018)
4. Bach, L.M., Mihaljevic, B., Zagar, M.: Comparative analysis of blockchain consensus algorithms. In: 2018 41st International Convention on Information and Communication Technology, Electronics and Microelectronics (MIPRO), pp. 1545–1550. IEEE (2018)
5. Buterin, V., et al.: A next-generation smart contract and decentralized application platform. White paper (2014)
6. Sovrin-Protocol and Token-White-Paper.pdf (2019). https://sovrin.org/wp-content/uploads/Sovrin-Protocol-and-Token-White-Paper.pdf. Accessed 17 Jan 2019
7. JP Morgan Chase. Quorum white paper (2016). https://github.com/jpmorganchase/quorum/blob/master/docs/quorum%20whitepaper%20v0.2.pdf. Accessed 17 Jan 2019
8. Hyperledger Whitepaper (2016). http://blockchainlab.com/pdf/hyperledger%20whitepaper.pdf. Accessed 17 Jan 2019
9. Bankchain.pdf (2019). http://www.bankchaintech.com/docs/brochures/bankchain.pdf. Accessed 17 Jan 2019
10. Odelu, V., Das, A.K., Rao, Y.S., Kumari, S., Khan, M.K., Choo, K.K.R.: Pairing-based CP-ABE with constant-size ciphertexts and secret keys for cloud environment. Comput. Stand. Interfaces **54**, 3–9 (2017)
11. Bao, F., Deng, R.H., Zhu, H.: Variations of Diffie-Hellman problem. In: International Conference on Information and Communications Security, pp. 301–312. Springer, Heidelberg (2003)
12. Dodis, Y., Ostrovsky, R., Reyzin, L., Smith, A.: Fuzzy extractors: how to generate strong keys from biometrics and other noisy data. SIAM J. Comput. **38**(1), 97–139 (2008)

13. Simoens, K., Bringer, J., Chabanne, H., Seys, S.: A framework for analyzing template security and privacy in biometric authentication systems. IEEE Trans. Inf. Forensics Secur. **7**(2), 833–841 (2012)
14. Patel, V.M., Ratha, N.K., Chellappa, R.: Cancelable biometrics: a review. IEEE Signal Process. Mag. **32**(5), 54–65 (2015)
15. Harn, L., Xu, Y.: Design of generalised ElGamal type digital signature schemes based on discrete logarithm. Electron. Lett. **30**(24), 2025–2026 (1994)
16. Odelu, V., Das, A.K., Wazid, M., Conti, M.: Provably secure authenticated key agreement scheme for smart grid. IEEE Trans. Smart Grid **9**(3), 1900–1910 (2018)

Lightning Network: A Comparative Review of Transaction Fees and Data Analysis

Nida Khan$^{(\boxtimes)}$ and Radu State

University of Luxembourg, 29 Avenue JF Kennedy,
Luxembourg City, Luxembourg
{nida.khan, radu.state}@uni.lu

Abstract. Blockchain is a revolutionary, immutable database disrupting the finance industry with a potential to provide payments in a secure environment, unhindered by intermediaries. However, scalability and throughput issues plague the technology and prevent it's mass scale adoption. The paper focusses on Lightning Network, the off-chain, scalable and high throughput payment solution from Bitcoin. A comparison is conducted to highlight the fee incurred for payment transactions through Lightning Network, Raiden, Stellar, Bitcoin and conventional payment systems to assess its viability as a blockchain-based payment system. The paper also provides an analysis of the data of Lightning Network, to give a global overview of its usage and reachability.

Keywords: Blockchain · Lightning Network · Data analysis · Transaction fees

1 Introduction

Blockchain is a ledger of transactions comprising of a peer to peer network and a decentralized distributed database. Bitcoin was the first blockchain platform to be launched. Other blockchains like Ethereum, Hyperledger and Stellar among others followed. There is ongoing research to solve the scalability and throughput issues related to blockchain platforms, while maintaining a decentralized infrastructure. Lightning Network was launched to solve the scalability and performance issues in Bitcoin and is an off-chain network that runs parallel to the Bitcoin blockchain. Bitcoin was conceptualized as a peer to peer payment network in a paper by Satoshi Nakamoto in 2008 [1]. At present Bitcoin supports 7 transactions per second [2] with a block size limit of 1 MB and a blockchain size of 235.29 GB [3]. If block size limit was increased to replace all other global financial transactions through Bitcoin, then the entire network would collapse or at the most lead to extreme centralization of Bitcoin nodes to the economically privileged. Further the storage requirements for the increased block size as well as the bandwidth requirements would be beyond the capabilities of home computers making Bitcoin lose it's utility for the masses. In order to scale the Bitcoin network, it was concluded that the transactions need to be off the Bitcoin blockchain [4]. Thus, the inception of Lightning Network took place to have a throughput of nearly unlimited number of transactions per second with very low fees.

Lightning Network was developed as a payment solution serving as an alternative to Bitcoin and seeks to cater to micropayments as well. Micropayments is a domain that

© Springer Nature Switzerland AG 2020
J. Prieto et al. (Eds.): BLOCKCHAIN 2019, AISC 1010, pp. 11–18, 2020.
https://doi.org/10.1007/978-3-030-23813-1_2

has not been exploited and still remains an area of untapped potential [5]. Lightning Network can serve as a payment solution for digital goods and services, where the costs are very low, ranging in the micropayments domain, and negligible transaction fees can be an advantage. It can also facilitate faster and cheaper cross-border transaction flows as compared to traditional payment methods. The paper is a pioneer in conducting both a comparative review of whether Lightning Network is a viable payment solution or not, using transaction fees as the evaluating parameter, and an analysis of its data to give an estimate of its usage and reachability. The paper gives the background and related work in Sect. 2. A comparison of the fee incurred for payment transactions through conventional and blockchain-based payment solutions is given in Sect. 3. Analysis of Lightning Network data is given in Sect. 4, while the conclusion is provided in Sect. 5.

2   Background and Related Work

Lightning Network is a second layer payment network developed on top of the Bitcoin blockchain platform. Theoretically it consists of an infinite number of bidirectional payment channels between users and can be used with other blockchain platforms too. A payment channel allows two transacting entities to do as many transactions as desired off-chain with only the initial and final transactions being recorded on the Bitcoin blockchain, incurring transaction fees. Hence, the fee for multiple off-chain transactions is the same as the fee for two transactions on Bitcoin. A payment channel is required for two entities to conduct payment transfers, the creation of which incurs high fee (Table 1). However, Lightning Network also facilitates payment transfers through an intermediary in the network, who has payment channels with the two entities. This feature is extended by incorporating multiple intermediaries in the network to conduct a payment transfer from one entity to another leading to a web of payment channels. The intermediary charges a very low fee for providing the payment channel (Table 1). The payment channels utilize multisignature [6] technology and locktime [7] to ensure a secure payment transaction without the need to trust the other party and the intermediaries, if involved. Lightning Network employs onion routing to securely and anonymously route payments within the network [8].

Sampolinsky and Zohar proposed the GHOST rule [9] offering performance benefits over the longest chain rule in Bitcoin. In [10], a brief overview of emerging directions in scalable blockchains is given with a discussion on the proof of work and Byzantine Fault Tolerant consensus mechanisms. Burchert *et al.* proposed an addition of a third layer to Bitcoin to function as an enhancement for the second layer, Lightning Network, as a means to bring about cost reduction and scalability [11]. Prihodko *et al.* proposed a new payment routing algorithm for Lightning Network [12]. Roos *et al.* proposed a decentralized routing for path-based transaction networks, like Bitcoin and Ethereum [13]. Pass and Shelat put forward a new lottery-based scheme for micropayments for ledger-based transaction systems [14]. Our work involves data analysis of Lightning Network and a comparative review of the transaction fees to evaluate it's potential to become a feasible payment solution.

3   Comparison of Transaction Fees

Lightning Network was released to serve as a scalable low cost payment solution but it provides less secure transactions than Bitcoin [4]. Hence, the following discussion evaluates the costs of payment transfers of small amounts of $1 from Alice to Bob and 50 cents from Alice to David through it. It compares the cost incurred with similar payment transfers through Raiden, Stellar, Bitcoin, MasterCard, Bank of America and PayPal. The fees are reflected in US dollars in Table 1 to facilitate an easy comparison between the different payment systems. The fee incurred in the relevant cryptocurrency is indicated alongside. The conversion rate used is applicable for a specific day [15]. The cryptocurrency value and the transaction fee in blockchains is volatile. Hence, the indicated costs in dollars would change accordingly.

3.1   Lightning Network

Lightning Network functions by registering the transactions to open and close a payment channel on the Bitcoin blockchain. Alice and Bob need to set up a payment channel between them to send some BTC, the cryptocurrency of Bitcoin. They deposit funds, $3 each in BTC, to open a channel and thereafter broadcast this deposition, which gets recorded on the Bitcoin blockchain. The payment transfers through the opened payment channel cannot exceed the deposited funds, which is referred to as the *channel capacity*. In our analysis, we consider the transaction fee needed to include the transaction in the next block of Bitcoin (10 min), which is 18 satoshis/byte (satoshi is the smallest unit of bitcoin cryptocurrency) [15]. Thereafter, Alice can send BTC equivalent to $1 to Bob accomplishing a direct payment transfer and close the payment channel. The procedure can be repeated in parallel with David to send 50 cents to him. The total fee and time for direct payment transfers in Table 1 indicates the costs and time for above. In the case of a payment transfer through an intermediary, let us assume that Bob already has a payment channel with David and Alice opens a payment channel with Bob. Alice sends $1 in BTC to Bob and 50 cents in BTC to David, through the payment channel of Bob paying the channel fee to him and closes the channel. The channel fee is 1 satoshi [16]. Table 1 gives the total fee and time for payment transfers from Alice to Bob and from Alice to David with Bob as intermediary in mediated payments. The lower bound of the transfer time is indicated and it can increase in periods of network congestion. The '+' sign used in the total time for payment transfers indicates a few seconds more. Similar payment transfers can be accomplished in Bitcoin through two transactions as seen in Table 1.

3.2   Raiden

Raiden Network [17] is the off-chain scaling solution for Ethereum blockchain [18] network. Ethereum provides the feature of smart contracts [19]. Raiden helps in instant, low fee payment transfers based on ERC20 tokens. ERC20 is a token standard, which describes the functions and events that an Ethereum token contract has to implement. The payment process is similar to Lightning Network and payment channel technology is employed to enable low cost, bidirectional payments. The transacting entity needs to have ERC20 tokens in an address, which needs to be registered with Raiden. Once registered by deploying a *Token Network Contract* on Ethereum, a token has a *Token Network* associated with it and the Token networks are responsible for opening new payment channels between transacting entities. If a transacting entity needs to send a payment transfer in an ERC20 token, which is already registered, then the costs for registering the token are absolved. The payment process after registration is akin to Lightning Network where the transacting entity needs to open a payment channel with another only if there are no intermediaries connecting them by depositing some tokens. As before the payment transfers through the channel cannot exceed the deposited tokens. The costs for token registration on Ethereum is 3.5 million gas [20], where gas is the unit to measure computational effort needed to execute an operation on the *Ethereum Virtual Machine* to calculate the costs in Ether (ETH), the cryptocurrency of Ethereum. We consider the fastest time the transaction can be included in Ethereum similar to our consideration for inclusion in the next block of Bitcoin for Lightning Network and it costs 0.091 ETH or $9.55 [21]. The mean time for transaction confirmation is presently 38 s [21] whereas theoretical limit was 14 s [2]. Data for channel fee is not available but it is predicted to be so low that the overall fee would not be affected significantly by its inclusion. The transaction cost for open and closing the payment channel in Ethereum for fastest transaction time with 21000 gas is 0.0005 ETH [21]. In Table 1 the costs and time for the given payment transfers from Alice for both unregistered and registered token are depicted. The methodology for fee computation for the payment transfers is the same as in Lightning Network. A mediated transfer includes the channel fee assumed as $0.

3.3   Stellar

Stellar is an open-source, distributed, blockchain-based payments infrastructure. Stellar aids in the optimum conversion of fiat currency into cryptocurrency, XLM, to enable fast cross-border payments between different currencies at extremely reduced rates between people, payment systems and financial organizations [22]. Sending a payment is an operation in Stellar and every operation has a base fee of 10^{-5} XLM [22]. The transaction fee depends upon the base fee and the number of operations. When a transacting entity conducts a payment transfer of some XLM to another, then a default fee is charged (Table 1). The default fee is independent of the amount transferred with the transfer time being 3–5 s [2]. Table 1 depicts the total fee and time for direct and mediated payments from Alice to Bob and David in Stellar.

3.4   Conventional Payment Methods

We consider payment transfers through a few popular payment solutions like PayPal, bank transfer through Bank of America and MasterCard. The transfer time varies from a few hours to several days in conventional payment methods. Since the amount is low, micropayment rates of PayPal would apply, which are 5% + $0.05 [23] of the paid amount. Payment transfers of $1 and 50 cents would therefore incur a total cost of $0.175. Bank of America charges $30 for a domestic wire transfer and more for international money transfers [24] so payments of $1 and 50 cents through a similar bank is not feasible. The credit card company MasterCard has a payment transaction fee of 0.19% + 0.53 [25] of the transacted amount in United States, which would cost a total fee of $1.063 for payment transfers of $1 and 50 cents.

Table 1. Total transaction fee and time for 2 payment transfers

Payment system	Lightning network	Raiden	Stellar	Bitcoin
Open channel: Fee, Time	$0.16 (18 satoshi/byte), 600 s	$0.05 (0.0005 ETH), 38 s	-	-
Close channel: Fee, Time	$0.16 (18 satoshi/byte), 600 s	$0.05 (0.0005 ETH), 38 s	-	-
Direct payment: Fee, Time	$0, milliseconds to seconds	$0, sub-seconds	$0.08 ($10^{-5}$ XLM), 3–5 s	$0.16 (18 satoshi/byte), 600 s
Channel: Fee, Time	$0.00003(1satoshi), few seconds	very low, few seconds	-	-
Total: Fee, Time- 1. Direct	$0.64, 1200 s+	$9.75[a], 114 s[a]+ $0.20[b], 76 s[b]+	$0.16, 3–5 s	$0.32, 600 s
2. Mediated	$0.32003, 1200 s+	$9.65[a], 114 s[a]+ $0.10[b], 76 s[b]+	$0.16, 3–5 s	-

[a]Unregistered token, cost for registration is 0.091 ETH.
[b]Registered token.

4   Data Analysis of Lightning Network

A beta version of Lightning Network was launched on the Bitcoin mainnet in March 2018 and we extracted data related to Lightning Network nodes, payment channels and channel capacity. Information was extracted concerning node ID, total number of payment channels of a node, total number of open channels and total number of closed channels. Our data collection dates till the first 5 months of its public release and reflects the state of the Lightning Network parameters till that time. In all, it was found that 60 countries have at least one Lightning Network node. We observed that US, which had 1141 nodes, out of the total 1983, owned 57.5% of the nodes. Germany ranked second with 165 nodes and France ranked third with 80 nodes. At the lower rung we had countries like Iceland, Malta, Peru with just one node and Indonesia, Thailand and Chile with 2 nodes.

Figure 1 represents the top 7 countries in decreasing order of the total number of Lightning Network (LN) nodes on the *x-axis*. The *y-axis* represents the channel capacity per node per country and is calculated by multiplying the total channel capacity of the country by a factor of $k = 1000$. The result is then divided by the total number of nodes in the country. The usage of the factor k is to normalize the data for optimum analysis and visualization. It was observed that US, which had the highest number of nodes, had the lowest channel capacity of 0.0362 BTC per node among the 7 countries. France ranked third in the total number of nodes but had the highest channel capacity of 0.0666 per node.

We also analyzed the total channel capacity of all the nodes found in 60 countries. The mean channel capacity was found to be 1.45, the standard deviation was 5.44 and the variance of the analyzed data was 29.64. The median was found to be at 0.24 BTC, which gives us the channel capacity lying in between the highest of 41.32 BTC of US and 0 BTC of Latvia. Argentina and Greece had a total channel capacity of 0.24 BTC. Three countries Uruguay, Latvia and Iceland had zero channel capacities with no open payment channels.

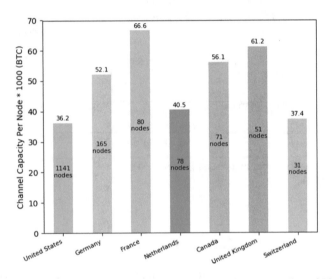

Fig. 1. Channel capacity per node in countries with the highest number of LN nodes

Figure 2 depicts 7 countries, with the highest number of open payment channels with labels on the *x-axis* giving the channel capacity per open channel in a country. Normalized data is represented in the labels and it needs to be divided by a factor of $k = 1000$ to derive the actual channel capacity. It was observed that UK had the highest channel capacity per open channel at 0.0102 BTC whereas US had the lowest at 0.0065 BTC. We also observed the total number of open and closed payment channels per country and it was seen that US has both the highest number of open payment channels at 6400 and the highest number of closed payment channels at 9279.

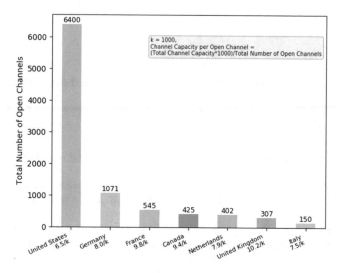

Fig. 2. Countries with the highest number of open channels

5   Conclusion

In this paper a comparison of the analyzed payment solutions indicates that the fee incurred for a payment transfer through Lightning Network is less than in Raiden (unregistered token), while Stellar provides the fastest payment transfer. Lightning Network and Raiden compute transaction fee based on the number of intermediaries, independent of the amount transferred. Stellar charges a default fee independent of both the payment amount and the number of hops in the network. It is also observed that the utility of Lightning Network lies in conducting multiple payment transfers using intermediaries and Bitcoin would be cheaper to use if only direct payment transfers are involved. PayPal comes close to offering similar transaction fees. Data analysis of Lightning Network reveals that United States is at the forefront of using the technology since it has the highest number of Lightning Network nodes, highest number of open channels and highest total channel capacity for payments. Lightning Network has the potential to become a viable payment solution catering more to the micropayment sector, as the channel capacity restricts the payment amount. Financial institutions as intermediaries to provide liquidity will help to strengthen it as a payment solution. Future work would involve analysis of the data of Raiden and Stellar to bring about an optimum assessment of the comparison between different blockchain-based payment solutions.

References

1. Nakamoto, S.: Bitcoin: A Peer-to-Peer Electronic Cash System. https://bitcoin.org/bitcoin.pdf
2. Lin, L.: Stellar-Uniqueness/ Differentiation – II. https://stellarcommunity.org/t/stellar-meetup-in-singapore/1665/2. Accessed 28 Jan 2019
3. BitInfoCharts, Cryptocurrency Statistics. https://bitinfocharts.com/. Accessed 28 Jan 2019
4. Poon, J., Dryja, T.: The Bitcoin Lightning Network: Scalable Off-Chain Instant Payments. https://lightning.network/lightning-network-paper.pdf
5. Ali, T., Clarke, D., McCorry, P.: The Nuts and Bolts of Micropayments: A Survey. CoRR (2017)
6. Bitcoin Wiki, Multisignature. https://en.bitcoin.it/wiki/Multisignature. Accessed 2 Jan 2019
7. Bitcoin.org, Bitcoin Developer Guide. https://bitcoin.org/en/developer-guide#term-locktime. Accessed 4 Jan 2019
8. GitHub Lightning Network, Onion Routed Micropayments for the Lightning Network. https://github.com/lightningnetwork/lightning-onion. Accessed 10 Jan 2019
9. Sampolinsky, Y., Zohar, A.: Secure high-rate transaction processing in bitcoin. In: Financial Cryptography and Data Security-19th International Conference (2015)
10. Vukolic, M.: The quest for scalable blockchain fabric: In: IFIP WG 11.4 Workshop, iNetSec (2015)
11. Burchert, C., Decker, C., Wattenhofer, R.: Scalable funding of bitcoin micropayment channel networks - regular submission. In: SSS (2017)
12. Prihodko, P., Zhigulin, S., Sahno, M., Ostrovskiy, A., Osuntokun, O.: Flare: An Approach to Routing in Lightning Network. White Paper (2016)
13. Roos, S., Moreno-Sanchez, P., Kate, A., Goldberg, I.: Settling payments fast and private: efficient decentralized routing for path-based transactions. In: NDSS (2018)
14. Pass, R., Shelat, A.: Micropayments for decentralized currencies. In: Proceedings of the 22nd ACM SIGSAC Conference on Computer and Communications Security (2015)
15. Bitcoin Transaction Fees. https://bitcoinfees.info/. Accessed 29 Jan 2019
16. Coindesk News, You Can Now Get Paid (a Little) for Using Bitcoin's Lightning Network. https://www.coindesk.com/you-can-now-get-paid-a-little-for-using-bitcoins-lightning-network. Accessed 28 Jan 2019
17. Raiden Homepage, The Raiden Network. https://raiden.network/. Accessed 28 Jan 2019
18. Wood, G.: Ethereum - A Secure Decentralized Generalized Transaction Ledger. https://gavwood.com/paper.pdf
19. Szabo, N.: Formalizing and securing relationships on public networks. First Monday 2(9) (1997)
20. Raiden GitHub, Getting Started with the Raiden API. https://raiden-network.readthedocs.io/en/stable/api_walkthrough.html
21. ETH Gas Station. https://ethgasstation.info/calculatorTxV.php. Accessed 29 Jan 2019
22. Stellar. https://www.stellar.org/developers/. Accessed 29 Jan 2019
23. PayPal. https://www.paypal.com/ca/webapps/mpp/merchant-fees. Accessed 29 Jan 2019
24. Bank of America. https://www.bankofamerica.com/foreign-exchange/wire-transfer.go. Accessed 29 Jan 2019
25. MasterCard. https://www.mastercard.us/content/dam/mccom/en-us/documents/merchant-interchange-rates.pdf. Accessed 29 Jan 2019

Do Smart Contract Languages Need to Be Turing Complete?

Marc Jansen, Farouk Hdhili, Ramy Gouiaa, and Ziyaad Qasem[✉]

Computer Science Institute, University of Applied Science Ruhr West,
Bottrop, Germany
{marc.jansen,ziyaad.qasem}@hs-ruhrwest.de

Abstract. Blockchain based systems become more and more prominent. While starting by developing (crypto)currency payment schemes, a lot of the latest development goes in the direction of executing source code directly in the peer-to-peer network blockchains are usually built on. These so called smart contracts have become popular in order to reduce the amount of necessary middle-mans involved in different processes. Despite the large amount of research already invested in the design of languages which support smart contracts, there are still a lot of problems in the existing approaches, regularly resulting in security flaws. One of these problems is the complexity of the used languages. Therefore, this paper provides an evaluation of currently deployed smart contracts with respect of the requirements of those contracts concerning computability. The finding is that most of the currently deployed smart contracts do not need Turing complete languages, but could also be implemented based on a simpler design of the underlaying language.

Keywords: Computability · Turing complete language · Blockchain

1 Introduction

In recent years, blockchain based systems gained a lot in popularity. Beside the most usual and first use case of a cryptocurrency, blockchains are nowadays discussed for a much broader amount of use cases, e.g., title registers [1], automatic contract execution [2] and so on. Therefore, additional requirements came up, especially the requirement to directly execute small programs on the blockchain, usually referred to as smart contracts. Here, smart contracts refer to distributed coordination task performed by nodes in the peer-to-peer network of a blockchain [3]. Smart contracts have especially become popular in order to remove middle-man activities in coordination tasks, while at the same time guarantee the consensus in the network. Due to the increased interest in this kind of automatic contract execution technology, a lot of research has already been invested in the design of languages that support the development of smart contracts. Still a lot of problems in the existing approaches, regularly resulting in security flaws, exist which lead to large, mostly financial, damage. One of

© Springer Nature Switzerland AG 2020
J. Prieto et al. (Eds.): BLOCKCHAIN 2019, AISC 1010, pp. 19–26, 2020.
https://doi.org/10.1007/978-3-030-23813-1_3

these problems is the complexity of the used languages. Therefore, one question currently discussed heavily in the blockchain communities is if programming languages with a lower complexity might be able to increase the security of those systems on the one hand and on the other hand if such programming languages are appropriate for the development of smart contracts. In order to answer one of those questions, this paper provides an evaluation of smart contracts on the Ethereum blockchain, as the currently largest blockchain network supporting those contracts. The major perspective of the presented research is to gather the requirements of those deployed smart contracts with respect to basic questions about computability theory. Therefore, a large number of smart contracts have been analyzed in order to identify if a Turing complete programming language is necessary for their implementation, or if less complex programming languages could be used.

The remainder of the paper is organized as follows: first the current research status with respect to the topic of this paper is presented in a state of the art section. Afterwards, the research question and analysis strategy used for the presented research is described, followed by a detailed description of the found results. Finally, the paper finishes by short discussion of the results and the presentation of possible future lines of research.

2   Background

After the introduction of the blockchain technology in 2008 [4], Bitcoin was the first implementation of this approach which went live in 2009. Ever since, a large number of different blockchain implementations appeared and all of them implemented a different notion of the proposed architecture. While Bitcoin itself is primarily meant to be a cryptocurrency, other implementations, e.g., Ethereum concentrated more on being a runtime environment for code executed directly on the blockchain, often referred to as chaincode or smart contracts.

Due to the high amount of values that are distributed by those blockchain implementations, they provide interesting targets to attack, partly with great success [5]. Therefore, this section provides a small overview of scripting approaches by different blockchains, allowing to implement code that is executable directly on the blockchain.

Sometimes not so well understood, every transaction in the cryptocurrency blockchain of Bitcoin is protected by a small script. This script is formulated in the Bitcoin Script language, stack based and attached to a transaction. If and only if the script attached to a transaction is successfully executed, the corresponding transaction is executed in the Bitcoin network, hence the corresponding funds are transferred from one to another address. Certain operations are possible in Bitcoin scripts like defining constants, doing basic flow control, splicing strings, bitwise logic, basic arithmetics and cryptographic functions. One important aspect about the Bitcoin scripting language is that although it does provide basic flow control mechanisms, e.g., conditions, it does not provide a Turing complete language, e.g., it does not provide means for loops and/or complex recursion.

Ethereum, as the currently second most prominent blockchain implementation, is an open-source distributed computing platform featuring smart contracts. It provides the ability to formulate smart contracts base on a language called Solidity. It provides an object oriented programming paradigm and made use of prominent ideas from JavaScript community, e.g., when it comes to syntax. While Solidity itself is a Turing complete programming language, the runtime environment in which Solidity based smart contracts are executed on the Ethereum blockchain limits the amount of computational power that each smart contract can consume, so that the developers of this approach speak of a pseudo-Turing complete approach. Users of the system are able to create their own token via smart contracts. Those tokens can basically be used for different purposes, e.g., represent shares of a company, could provide access to certain services or act as an investment fund. One of the advantages of this approach is, that developers of smart contracts are very flexible with respect of the functionality implemented in the smart contracts. On the other hand, the more complex the implementation of smart contracts are, the more error prone they are at the same time. A couple of prominent hacks on the Ethereum blockchain are due to errors in the complex implementation of their corresponding smart contracts, e.g., the DAO hack [6].

In contrast to Bitcoin and Ethereum, NEO did not introduce a new programming language for the formulation of smart contracts, but it targets on the integration of well-known, respected and widely spread existing programming languages in which smart contracts for the NEO platform should be implemented. The major idea of this approach is on the one hand to open the blockchain world for a larger community by offering interfaces to blockchains based on well-known programming languages so that developers do not need to learn new programming languages in order to interact with the technology. On the other hand, this approach allows to some extent to mitigate the risks involved with the developed smart contracts. As the developer uses his/her usual IDE debugging tool, it is less likely that there will be bugs that could affect the contract behavior at runtime.

In contrast to the approach of Ethereum or NEO to provide Turing complete languages for the development of smart contracts, the Waves Platform decided to take a two steps approach. In the first step a non-Turing complete programming language, called RIDE, was developed in order to provide basic scripting functionality, while at the same time to reduce the risk of security relevant errors in those scripts. The RIDE language does, e.g., not support recursion or any kind of loops. This, on the one hand, makes the programming structure very simple so that scripts could easily be tested against security flaws, while at the same time it allows to estimate the computational effort necessary for the execution of the smart contract, e.g., for calculating costs associated with the execution and/or ensuring that there are no smart contracts deployed that allow for denial of service attacks against nodes of the blockchain. In general it could be said, that the halting problem [7] is easy to decide for scripts implemented in RIDE.

Indeed, there is an active discussion going on with respect to the question if the scripting language in Bitcoin is actually Turing complete. There is a bit of research already available on that topic, e.g., [8,9] and [10]. Basically, the idea here is that the forging/mining mechanism that generates new blocks, could also be seen as an endless loop that could, together with conditions, be used in order to provide the functionality of a while loop, allowing while-computable calculations and thus providing a Turing complete approach.

The next section describes the research question of the presented work along with the analysis strategy that was implemented in order to analyze the current situation with respect to the research question.

3   Research Question and Analysis Strategy

The research question driving the motivation of the presented research was to identify the need of Turing complete languages for the implementation of smart contracts in blockchain based systems. As discussed in Sect. 2, different approaches exist, which either use Turing complete languages (like Ethereum and NEO) while other approaches intentionally use non-Turing complete languages for the implementation of smart contracts, e.g., Bitcoin and Waves. This question is particularly interesting having in mind that most attacks on blockchain based systems, have either been due to stolen passwords, server misconfigurations or attack vectors against poorly implemented smart contracts. Therefore, a prominent idea in the blockchain community is to provide less complicated programming languages in order to make errors in smart contract codes less probable and increase potential security checks on smaller and less complicated implementations. In this context, we can refer to the programming language Vyper[1]. Vyper is a python-like scripting language that targets the Ethereum Virtual Machine (EVM). The goal of this language is to write secure and maximally human-readable smart contacts by losing some Turing complete features. This shows that the developers have indeed begun to think about solutions to the problems of smart contracts caused by the Turing complete languages. At the same time, a non-Turing complete programming language for the implementation of smart contracts would definitely limit the type of contracts that could be implemented.

Yet often mentioned scenarios as examples from the blockchain field as Atomic Swaps, securing funds in crowdfunding campaigns and further more could also be implemented based on non-Turing complete programming languages. In this context, Atomic Swaps [11] refer to sending tokens from one blockchain based implementation to another one in which either the corresponding transactions on both sides are executed or none of them. The other example of securing funds during a crowdfunding campaign is reasonable in order to prevent funds being transferred from a certain address before a certain point in time. This is a very usual demand during crowdfunding campaigns, e.g., in order to prevent someone from spending the funds before the campaign is finished. Of

---

[1] https://vyper.readthedocs.io.

course this concept of putting time restrictions on spending funds could more generalized also be used in other scenarios. Therefore, since there is a significant amount of examples that do not rely on Turing complete programming languages, it is fair to raise the question on how big the percentage of contracts is, that actually need Turing complete programming languages. In order to further analyze this, the smart contracts currently deployed to the Ethereum blockchain seem to be a good point of reference, since the Ethereum platform is currently by far the largest platform that allows for the deployment of Turing complete smart contracts.

The next section describes the implementation of the different steps in order to perform the described analysis of Ethereum based smart contracts with respect to their demand for Turing complete languages.

4   Implementation

As described in the last section, the goal of this research is to answer the question if Turing complete languages are necessary for smart contracts. In order to do this, the verified smart contracts on the Ethereum blockchain are analyzed with respect to different paradigms used in the contracts, e.g., like loops and/or recursion.

4.1   Dataset

Smart contracts as code that could be executed on the blockchain directly offer functions that can be used by participants all over the world and they are accessible to everyone, making them a valuable target for hackers. Therefore, a formal verification process has been deployed on the Etherscan platform. Each Solidity smart contract code is compiled to EVM Bytecode that runs on the Ethereum network which means each Solidity script has its own Creation Address Bytecode. A hacker or even malicious miner may try to cheat by not running the program or running it incorrectly [12]. To verify a smart contract, the developer has to apply a proof that the contract behaves as intended, in order to reduce the risk of malicious scripts, otherwise if the Bytecode generated matches the existing Creation Address Bytecode, the contract is verified.

Unfortunately, the publicly available API of the platform does not provide access to smart contracts. Therefore, a web scraping tool [13] was implemented allowing to extract data from one of the websites that provide access to the smart contracts. Etherscan.io[2] provides a central starting point for accessing verified smart contracts on the Ethereum blockchain using the web scraping tool. Additionally, Etherscan.io also provides access to specific verified smart contracts in order to access certain details of a particular contract.

In order to actually download and store the available verified smart contracts from Etherscan, a NodeJS program with the cherioo web scraping module was

---

[2] https://etherscan.io/.

implemented. By utilizing this program all available verified smart contracts on the Ethereum blockchain at the point of writing have been downloaded and stored. This results in a total of 53757 verified smart contracts that have been made available for later analysis, providing a rich enough base for the conclusions drawn by the results presented later in this paper.

4.2 Analysis of the Smart Contracts

Based on the usual classification of different computability classes, as shown in Fig. 1, the stored smart contracts have been analyzed with respect to the flow control mechanisms they provide.

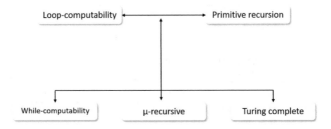

Fig. 1. Usual differentiation of computability classes

According to the well-known theory, while-computability is equivalent to Turing completeness and to μ-recursive functions. At the same time, loop-computability is equivalent to primitive recursion, while both computability classes are less powerful in comparison to while-computability, μ-recursive functions or Turing completeness, since there exist functions from the latter computability classes that are not loop-computable or could not be expressed with primitive recursion, while all loop-computable/primitive recursive functions are while-computable/μ-recursive/Turing complete. Therefore, the performed analysis identified the flow control mechanisms used by the extracted smart contracts and assigned each smart contract to a corresponding computability class in order to check for the percentage of smart contracts that actually need a Turing complete smart contract for their implementation and by this answer the raised research question. Regular expressions [14] provide a powerful tool for the analysis of texts. Therefore, the source codes of the stored smart contracts have been interpreted as text files and been analyzed with regular expressions in order to check for different control flow mechanisms in the contract's codes.

5   Evaluation Results

Based on the mentioned analysis steps in the previous sections, we have found that about 24.8% (13347 of the analyzed 53757 contracts) of all verified contracts use for-loops. It is important to stress here, that this kind of contracts do strictly speaking not fall in the computability class that demands for Turing complete smart contracts. With respect to the problems related to the halting problem raised by these kinds of loops in blockchains, we considered these kinds of contracts also to the same complexity class.

Furthermore, we have found that about 3.6% make use of recursive functions. It is important to note once again that this type of contracts do not strictly speaking fall into the class of computability that requires Turing complete, but for the same practical reasons mentioned before, we counted them to the more complex computability class.

Related to the while-loops, about 6.9% of all analyzed smart contracts make use of while-loops. Apparently, this result was a bit surprising since strictly spoken, only those 6.9% of all analyzed smart contracts made use of a control flow mechanism that usually demands for a Turing complete programming language. In addition, it is highly probable that some of those smart contracts could be implemented without while-loops (can be implemented with for-loops with a variable amount of runs).

5.1   Discussion of the Results

The obtained results show that only a very small amount of currently verified smart contracts on the Ethereum blockchain (6.9% of while-loops) really fall into the complexity class of Turing complete functions. In other words, only 6.9% of all currently deployed smart contracts use problematic control-flow mechanisms.

Even if we take a more practical perspective, namely the for-loops and primitive recursive functions and the problems related to the halting problem associated with this type of function, we got only 35.3% (6.9% of while-loops, 24.8% of for-loop and 3.6% of contracts that use recursion). In such a case, 35.3% of all currently deployed smart contracts use problematic control-flow mechanisms. Therefore, the results imply that making use of non-Turing complete smart contracts in the context of blockchain based systems makes perfect sense. Nevertheless, it might be reasonable to provide a twofold approach in which one non-Turing complete language for smart contracts is provided, alongside a second programming language that allows for the execution of Turing complete smart contracts.

6   Outlook and Future Work

The presented work provides an answer to the research question if programming languages for smart contracts need to be Turing complete. It therefore analyzed verified smart contracts on the Ethereum blockchain with respect to the complexity of flow-control mechanisms used in those contracts. The results show that by far the larger amount of smart contracts do not need Turing complete languages for their formulation. Future steps to extend the results of this research might include a deeper analysis in two dimensions: first of all, the anomaly of not analyzing indirect recursive method calls could be analyzed further; second, the identified while-loops could be analyzed for possible implementations with other flow-control mechanisms. Nevertheless, the results presented in this research might have an important impact on the security of smart contracts based on blockchain technology, if developers of those smart contracts will be provided by non-Turing complete programming languages for the implementation of their smart contracts.

References

1. Tama, B.A., Kweka, B.J., Park, Y., Rhee, K.H.: A critical review of blockchain and its current applications. In: Proceedings of the International Conference on Electrical Engineering and Computer Science (ICECOS), pp. 109–113. IEEE (2017)
2. Wright, C., Serguieva, A.: Sustainable blockchain-enabled services: smart contracts. In: Proceedings of the International Conference on Big Data, pp. 4255–4264. IEEE (2017)
3. Alharby, M., Van Moorsel, A.: Blockchain-based smart contracts: a systematic mapping study (2017). arXiv preprint arXiv:1710.06372
4. Nakamoto, S.: Bitcoin: a Peer-to-peer Electronic Cash System (2008). https://bitcoin.org/bitcoin.pdf
5. Atzei, N., Bartoletti, M., Cimoli, T.: A survey of attacks on Ethereum smart contracts (SoK). In: Principles of Security and Trust, pp. 164–186. Springer, Heidelberg (2017)
6. Mehar, M.I., Shier, C.L., Giambattista, A., Gong, E., Fletcher, G., et al.: Understanding a revolutionary and flawed grand experiment in blockchain: the DAO attack. SSRN Electron. J. (2017). https://doi.org/10.2139/ssrn.3014782
7. Church, A.: An unsolvable problem of elementary number theory. Am. J. Math. **58**(2), 345–363 (1936)
8. Wright, C.: Turing Complete Bitcoin Script White Paper (2016). https://ssrn.com/abstract=3160279
9. Wright, C.: Beyond Godel (2018). https://ssrn.com/abstract=3147440
10. Sgantzos, K.: Implementing a church-turing-deutsch principle machine on a Blockchain. Department of Computer Science and Biomedical Informatics, University of Thessaly, Lamia, Greece (2017)
11. Herlihy, M.: Atomic cross-chain swaps (2018). arXiv preprint arXiv:1801.09515
12. Bhargavan, K., Delignat-Lavaud, A., Fournet, C., Gollamudi, A., Gonthier, G., et al.: Formal verification of smart contracts: short paper. In: Proceedings of the Workshop on Programming Languages and Analysis for Security, pp. 91–96. ACM, October 2016
13. Castrillo-Fernández, O.: Web Scraping: Applications and Tools (2005). https://www.europeandataportal.eu/sites/default/files/2015_web_scraping_applications_and_tools.pdf
14. Goyvaerts, J., Levithan, S.: Regular Expressions Cookbook. O'reilly, Sebastopol (2012)

Towards Integration of Blockchain and IoT: A Bibliometric Analysis of State-of-the-Art

Mohammad Dabbagh$^{(\boxtimes)}$, Mohsen Kakavand, and Mohammad Tahir

School of Science and Technology, Sunway University,
47500 Selangor, Malaysia
{mdabbagh,mohsenk,tahir}@sunway.edu.my

Abstract. Since its inception, Blockchain has proven itself as an emerging technology that revolutionizes diverse industries. Among others, Internet of Things (IoT) is one of the application domains that reaps large benefits from Blockchain. The Blockchain's potential to overcome different challenges of IoT services has shifted the research interests of many scientists towards addressing the integration of two disruptive technologies, i.e., IoT and Blockchain. This resulted in publishing more research papers in this emerging field. Thus, there is a need to conduct research studies through which a broad overview of research contributions in this field could be investigated. To respond to this need, a number of review papers have been published recently, each of which has considered the integration of IoT and Blockchain from a different perspective. Nonetheless, none of them has reported a bibliometric analysis of the state-of-the-art in the integration of IoT and Blockchain. This gap stimulated us to investigate a thorough analysis of the current body of knowledge in this field, through a bibliometric study. In this paper, we conducted a bibliometric analysis on the Scopus database to assess all scientific papers that addressed the integration of IoT and Blockchain. We have analyzed those collected papers against four criteria including annual publication and citation patterns, most-cited papers, most frequently used keywords, and most popular publication venues. The results disseminate invaluable insights to the researchers before establishing a research project on IoT and Blockchain integration.

Keywords: Blockchain · Internet of Things, IoT · Bibliometric study · Scopus

1 Introduction

The Internet of Things (IoT) has been revolutionizing the way people interact with the environment by connecting physical objects together, thus, transforming the real world into a huge information system. The IoT has established itself in the central process of diverting ordinary cities into smart cities, typical cars into connected cars and normal houses into smart homes. The number of IoT devices is growing at an astounding pace and several reports anticipated that the IoT devices would exceed 20 billion by 2020 [1]. As predicted by Gartner, IoT will encompass the core technology of 95% of all new devices by 2020 [2].

Every IoT device is able to connect to a network and exchange data over the Internet. Therefore, each IoT device generates a large amount of data, which could be processed

© Springer Nature Switzerland AG 2020
J. Prieto et al. (Eds.): BLOCKCHAIN 2019, AISC 1010, pp. 27–35, 2020.
https://doi.org/10.1007/978-3-030-23813-1_4

and analyzed in order to provide insights and knowledge, thereby, facilitating to make smarter decisions. Accordingly, proper architectures, protocols and standards have to be deployed in order to support IoT solutions. The prevalent remedy to accommodate the needs of IoT services is to take advantage of cloud computing technology. Although the integration of IoT and cloud computing has been promising over the years, the centralized architecture of cloud computing may lead to some significant issues like the reliability of data generated by IoT devices. How we can make sure that an intermediary party has not modified the information provided by IoT services. This brings out the need for researchers to explore a solution that has the potential to guarantee the reliability of data.

Blockchain technology has recognized itself as a prominent solution to overcome not only the reliability issue of information generated by IoT services, but also the security, privacy, and scalability issues of IoT solutions [3, 4]. Blockchain technology is able to perform, store, and keep track of transactions transparently on decentralized network infrastructure, keeping data immutable, reliable, and verifiable [5]. Therefore, many enterprises are experiencing the integration of two disruptive technologies, i.e., IoT and Blockchain, to improve their IoT services. Similarly, over the recent few years, many researchers have shifted their research interests towards addressing the integration of Blockchain and IoT. Consequently, we can observe an incremental trend in the number of research studies in this emerging field. This has created an opportunity for the researchers to conduct surveys through which a broad overview of research contributions in this field could be investigated. To respond to this need, a number of surveys have been published [6–9], each of which has considered the integration of IoT and Blockchain from a different perspective. Nonetheless, none of the current studies has reported a bibliometric analysis of the state-of-the-art in the integration of IoT and Blockchain. This gap stimulated us to investigate a thorough analysis of the current body of knowledge in this emerging field of research, through a bibliometric study.

The key objective of the current paper is to systematically collect, characterize and analyze all research papers that have addressed the integration of IoT and Blockchain. To attain the desired objective, we conducted a bibliometric study on the Scopus database to assess all scientific papers related to IoT and Blockchain integration with the aim of disseminating invaluable insights to the research community. To the best of the authors' knowledge, this paper delineates the first endeavour in the literature towards conducting such a bibliometric study. The results extracted from the bibliometric study presented in this paper would reveal (i) yearly publication and citation patterns; (ii) top-ten influential papers; (iii) most frequently used keywords; (iv) most popular publication venues that addressed the integration of IoT and Blockchain.

2 Research Questions and Dataset

RQ1: What is the evolution of IoT and Blockchain integration in terms of publication and citation trends?
RQ2: What are the most-cited papers related to IoT and Blockchain integration?
RQ3: What is the distribution of frequent keywords in related papers?
RQ4: What are the most popular publication venues for papers, which addressed IoT and Blockchain integration?

In this study, we chose Scopus as a data source and then started the data collection process by choosing *"blockchain" AND "internet of things"* as one single query string. The query targeted every article's title, abstract, and keywords and covered the time span of 2013 to the end of 2018. It should be highlighted that the data collection process was carried out on January 3, 2019. As a result, 451 papers, including 258 conference papers, 98 articles, 47 conference reviews, 31 articles in press, 10 reviews, 4 book chapters, 2 editorials, and 1 book, were retrieved to construct our dataset.

3   Results and Discussion

3.1   RQ1: What Is the Evolution of IoT and Blockchain Integration in Terms of Publication and Citation Trends?

Figures 1 and 2 represent the publication and citation trends of papers that addressed the integration of Blockchain and IoT in recent years. From Fig. 1, we can observe that before the year 2015 there has not been any research paper that focused on the integration of IoT and Blockchain. However, we can see the initial appearance of interest to this field by publishing 2 and 11 papers in 2015 and 2016 respectively. This trend has continued exceedingly in recent two years, i.e., 2017 and 2018, by publishing 104 and 334 papers in each respective year. This brings in a clear message to the research community that integration of IoT and Blockchain has opened up a new line of research in recent years, and it requires more attention from the researchers to address the gaps in this emerging field in the forthcoming years.

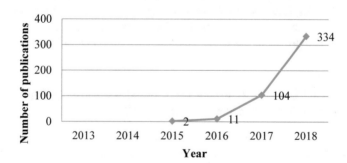

Fig. 1.   Yearly distribution of papers indexed by Scopus

Citation analysis is a useful way to evaluate how a particular field of research has impressed the research community. This motivated us to explore the yearly citation trends of papers that addressed the integration IoT and Blockchain. The results are sketched in Fig. 2 where we can see a significant change in the number of citations in 2017 (#157) compared to the previous years, i.e., 2016 (#7) and 2015 (#1). However, the number of citations is quite remarkable in 2018 (#1266) which confirms the importance level of this field for the research community. On the other hand, analysis of these results can provide some insights to the young researchers who are in the stage of initiating a research project.

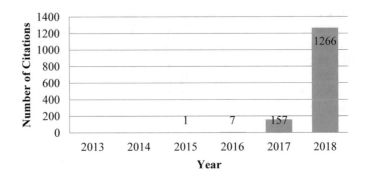

Fig. 2. Annual citations trend of papers related to IoT and Blockchain integration

3.2 RQ2: What Are the Most-Cited Papers Related to IoT and Blockchain Integration?

Table 1 indicates a shortlist of the ten most-cited papers that considered the integration of IoT and Blockchain. These papers are sorted based on the total number of citations (as shown in the rightest column of Table 1). Among others, the paper titled "Blockchains and Smart Contracts for the Internet of Things" [11] has recognized itself as the most-cited paper with 277 citations (to the date of conducting this research). The IEEE Access journal has published this paper in 2016. Table 1 also shows other useful information such as authors' name, the institution and country of the first author of each most-cited paper. This information can help researchers to better identify the possible choices for future research collaboration in this field. Furthermore, reviewing the highly cited papers would familiarize the researchers with the hottest and most influential research topics in this field, which can be helpful in initiating a new research project.

3.3 RQ3: What Is the Distribution of Frequent Keywords in Papers Related to IoT and Blockchain Integration?

To find out which topics have stronger links with IoT and Blockchain integration, we analyzed the number of occurrences of a keyword across all retrieved papers using VOSviewer. Initially, we set the minimum number of occurrences of a keyword to five in order to shortlist the number of keywords. Therefore, out of 950 keywords, 41 keywords met the threshold (As shown in Table 2). Then, for each of the 41 keywords, the total strength of the co-occurrences links with other keywords was calculated. Table 2 lists each selected keyword along with its respective occurrences as well as total link strength. Obviously, Blockchain and IoT are the most frequent keywords with highest total link strength, even though same keywords with different spelling are calculated separately, e.g., "iot", "internet of things (iot)", "internet-of-things", "blockchains", "blockchain technology".

Table 1. Overview of ten most-cited papers related to IoT and Blockchain integration

Title	Source	Country	Institution (of first author)	Year	Citations
Blockchains and Smart Contracts for the Internet of Things [10]	IEEE Access	USA	North Carolina State University	2016	277
Blockchain for IoT security and privacy: The case study of a smart home [11]	IEEE International Conference on Pervasive Computing and Communications	Australia	University of New South Wales	2017	73
An Overview of Blockchain Technology: Architecture, Consensus, and Future Trends [8]	IEEE 6th International Congress on Big Data, BigData Congress	China	Sun Yat-sen University	2017	56
Towards an optimized blockchain for IoT [12]	IEEE/ACM 2nd International Conference on Internet-of-Things Design and Implementation	Australia	University of New South Wales	2017	45
Internet of Things, Blockchain and Shared Economy Applications [13]	Procedia Computer Science	United Kingdom	University of Sussex	2016	44
Managing IoT devices using blockchain platform [14]	International Conference on Advanced Communication Technology, ICACT	South Korea	Electronics and Telecommunications Research Institute	2017	41
Can Blockchain Strengthen the Internet of Things? [15]	IT Professional	USA	University of North Carolina	2017	39
Securing smart cities using blockchain technology [16]	IEEE International Conference on High Performance Computing and Communications	Australia	Griffith University	2017	35

(*continued*)

Table 1. (*continued*)

Title	Source	Country	Institution (of first author)	Year	Citations
IoT security: Review, blockchain solutions, and open challenges [17]	Future Generation Computer Systems	Pakistan	Bahauddin Zakariya University	2018	33
The IoT electric business model: Using blockchain technology for the internet of things [18]	Peer-to-Peer Networking and Applications	China	Tsinghua University	2017	33

Figure 3 visualizes a cloud of keywords in all retrieved papers related to the integration of IoT and Blockchain. Important keywords are labeled with larger fonts, whereas different colors highlight different clusters. Links among keywords illustrate the strength level of co-occurrences between keywords. Some insights can be obtained by having a closer look at Table 2 and Fig. 3. We can see that security and privacy are two important topics that have been widely considered in the integration of IoT and Blockchain. Smart contracts and ethereum are the other two topics, which have a strong link with IoT and Blockchain integration. However, healthcare and smart home are the two domains that might investigate the integration of IoT and Blockchain in close future more than before. Exploring the applicability of machine learning algorithms with IoT and Blockchain would be also an interesting topic for future research.

3.4   RQ4: What Are the Most Popular Publication Venues for Papers, Which Addressed IoT and Blockchain Integration?

One of the most crucial parts of conducting any research is to target a suitable publication venue for presenting the outcome of a research study. This decision may also influence the citations number of a particular research paper after it is published. Thus, it is important to distinguish which publication venues have been good choices for publishing the outcomes of those researches that address the integration of IoT and Blockchain. Table 3 provides us with such information. Based on the information given in Table 3, Lecture Notes in Computer Science (including subseries Lecture Notes in Artificial Intelligence and Lecture Notes in Bioinformatics), ACM International Conference Proceeding Series, and IEEE Internet of Things Journal are the most popular venues based on the number of papers they published. Among journals, IEEE Internet of Things Journal, Sensors, and IEEE Access are the top three targeted publication venues for this emerging field of research. However, IEEE Access, Procedia Computer Science, and IEEE Internet of Things Journal are pioneers in terms of the number of citations.

Table 2. List of selected keywords along with the number of occurrences and Total Link Strength (TLS)

Keyword	Occurrences	TLS	Keyword	Occurrences	TLS
blockchain	279	540	smart home	8	22
internet of things	138	314	decentralized	6	20
iot	87	209	big data	6	18
security	43	139	smart grid	6	18
privacy	35	115	cryptocurrency	6	17
ethereum	30	82	distributed ledger	6	17
access control	19	66	identity management	5	17
smart contract	26	66	internet-of-things	6	17
smart contracts	25	64	sharing economy	5	17
bitcoin	19	50	data integrity	5	16
edge computing	15	42	blockchains	6	15
cloud computing	12	40	distributed systems	7	15
cybersecurity	9	35	block chain	6	14
internet of things (iot)	16	35	blockchain technology	12	13
trust	11	34	data security	5	13
authentication	10	32	consensus	6	12
supply chain	11	32	industry 4.0	5	12
fog computing	6	27	iot security	6	9
smart city	9	27	machine learning	5	8
cryptography	5	23	supply chain management	5	8
healthcare	7	22			

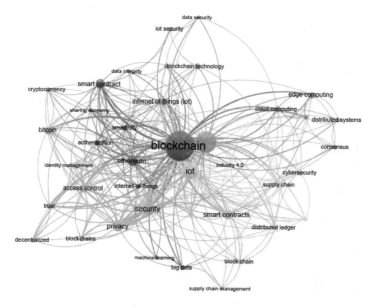

Fig. 3. Word cloud of frequent keywords in IoT and Blockchain integration

Table 3. Most popular venues of papers related to IoT and Blockchain integration

Venue	Publications	Citations
Lecture Notes in Computer Science	41 (#1)	5 (#6)
ACM International Conference Proceeding Series	24 (#2)	31 (#4)
IEEE Internet Of Things Journal	16 (#3)	36 (#3)
Sensors Switzerland	11 (#4)	26 (#5)
IEEE Access	9 (#5)	324(#1)
Procedia Computer Science	8 (#6)	53(#2)

4   Conclusions and Future Work

This research aimed at performing a bibliometric analysis of the state-of-the-art in the integration of IoT and Blockchain. To this end, we retrieved 451 papers from Scopus literature database with the time coverage of 2013 to the end of 2018. Publication and citation analyses of collected papers implied that integration of IoT and Blockchain has opened up new lines of research in recent years, especially in 2017 and 2018, and its impact has significantly impressed the research community. It is predictable that in the oncoming years the researchers would make more contributions to address the gaps in this emerging field. Detailed investigation of most frequent keywords hinted that security and privacy are two important topics that have been widely considered in the integration of IoT and Blockchain. IEEE Internet of Things Journal, Sensors, and IEEE Access have recognized themselves as the top three targeted journals for publishing research findings in IoT and Blockchain integration. As future work, it is worth exploring (i) the applicability of machine learning algorithms while integrating the IoT and Blockchain; and (ii) the integration of IoT and Blockchain on domains such as smart home, healthcare and supply chain management.

References

1. Díaz, M., Martín, C., Rubio, B.: State-of-the-art, challenges, and open issues in the integration of internet of things and cloud computing. J. Netw. Comput. Appl. **67**, 99–117 (2016)
2. Gartner. https://gtnr.it/2YSCkBw. Accessed 30 Mar 2018
3. Reyna, A., Martín, C., Chen, J., Soler, E., Díaz, M.: On blockchain and its integration with IoT. Challenges and opportunities. Future Gener. Comput. Syst. **88**, 173–190 (2018)
4. Casado-Vara, R., de la Prieta, F., Prieto, J., Corchado, J. M.: Blockchain framework for IoT data quality via edge computing. In: 1st Workshop on Blockchain-Enabled Networked Sensor Systems, pp. 19–24. ACM (2018)
5. Dabbagh, M., Sookhak, M., Safa, N.S.: The evolution of blockchain: a bibliometric study. IEEE Access **7**, 19212–19221 (2019)
6. Panarello, A., Tapas, N., Merlino, G., Longo, F., Puliafito, A.: Blockchain and IoT integration: a systematic survey. Sensors **18**(8), 2575 (2018)
7. Fernández-Caramés, T.M., Fraga-Lamas, P.: A review on the use of blockchain for the internet of things. IEEE Access **6**, 32979–33001 (2018)

8. Zheng, Z., Xie, S., Dai, H., Chen, X., Wang, H.: An overview of blockchain technology: architecture, consensus, and future trends. In: International Congress on Big Data, pp. 557–564. IEEE (2017)
9. Zheng, Z., Xie, S., Dai, H.N., Chen, X., Wang, H.: Blockchain challenges and opportunities: A survey. Int. J. Web Grid Serv. **14**(4), 352–375 (2018)
10. Christidis, K., Devetsikiotis, M.: Blockchains and smart contracts for the internet of things. IEEE Access **4**, 2292–2303 (2016)
11. Dorri, A., Kanhere, S.S., Jurdak, R., Gauravaram, P.: Blockchain for IoT security and privacy: the case study of a smart home. In: International Conference on Pervasive Computing and Communications Workshops, pp. 618–623. IEEE (2017)
12. Dorri, A., Kanhere, S.S., Jurdak, R.: Towards an optimized blockchain for IoT. In: 2nd International Conference on IoT Design and Implementation, pp. 173–178. ACM (2017)
13. Huckle, S., Bhattacharya, R., White, M., Beloff, N.: Internet of things, blockchain and shared economy applications. Procedia Comput. Sci. **98**, 461–466 (2016)
14. Huh, S., Cho, S., Kim, S.: Managing IoT devices using blockchain platform. In: International Conference on Advanced Communication Technology, pp. 464–467. IEEE (2017)
15. Kshetri, N.: Can blockchain strengthen the internet of things? IT Prof. **19**(4), 68–72 (2017)
16. Biswas, K., Muthukkumarasamy, V.: Securing smart cities using blockchain technology. In: 14th International Conference on Smart City, pp. 1392–1393. IEEE (2016)
17. Khan, M.A., Salah, K.: IoT security: review, blockchain solutions, and open challenges. Future Gener. Comput. Syst. **82**, 395–411 (2018)
18. Zhang, Y., Wen, J.: The IoT electric business model: using blockchain technology for the internet of things. Peer-to-Peer Netw. Appl. **10**(4), 983–994 (2017)

ClinicAppChain: A Low-Cost Blockchain Hyperledger Solution for Healthcare

Daniel-Jesus Munoz[1(✉)], Denisa-Andreea Constantinescu[2], Rafael Asenjo[2], and Lidia Fuentes[1]

[1] CAOSD Group, Universidad de Málaga,
29010 Málaga, Spain
{danimg,lff}@lcc.uma.es
[2] Computer Architecture Department, Universidad de Málaga,
29010 Málaga, Spain
{dencon,asenjo}@uma.es

Abstract. Unified access to anonymous records through a trustworthy system is needed in our increasingly globalised world that still suffers from profoundly disconnected health care services. Emergencies are followed by redundant, costly and slow medical examinations without a global health data sharing mechanism. We establish the foundations of a decentralised healthcare ledger where patients decide what to share, with who, and with minimal costs. We review the state-of-the-art of transparent, auditable and interactive Blockchain-based healthcare studies, developing ClinicAppChain. Our solution features authentication, confidentiality and permissioned data sharing, considering the EU data protection regulation of 2018. ClinicAppChain is a cross-platform low-cost Blockchain prototype that empowers patients, hospitals, researchers, pharmaceuticals and insurance industries without crypto-currencies involved, and with a negligible energy foot-print (7.7 watts per node).

Keywords: Blockchain · Hyperledger · Healthcare · Low-cost · Low-power

1 Introduction

Healthcare is a field that requires an efficient and secure system for managing medical records and tracking complex worldwide transactions. In this paper, we explore the feasability of a Borderless *Electronic Health Records* (EHR) Management System using blockchain technology. Blockchain is a distributed ledger of events, transactions or electronic records, which are cryptographically secured, impossible to clone, and can evolve through a consensus protocol. Each connected node works with that consensus [14]. Despite the popular focus on financial services, Blockchain has the potential to reshape how healthcare systems currently operate, empowering users (e.g., patients, practitioners, researchers) and fomenting improved medication, disease prevention, treatments and lifestyle tracking.

© Springer Nature Switzerland AG 2020
J. Prieto et al. (Eds.): BLOCKCHAIN 2019, AISC 1010, pp. 36–44, 2020.
https://doi.org/10.1007/978-3-030-23813-1_5

Depending on the consensus algorithm, there could be a tremendous energy foot-print of high-power demanding computing systems and scalability issues [11]. For instance, the culprit of the Bitcoin power demand is its consensus algorithm, *Proof of Work* (PoW) [15]. *Proof of Stake* (PoS) algorithm came as an alternative PoW, replacing the intensive computational work with a random selection process. It can be used both in public and private blockchains. One of the most successful blockchain platforms, Ethereum, plans to migrate from PoW to PoS. Voting-based or consensus algorithms as *Practical Byzantine Fault Tolerance* (PBFT) are preferred in private and semi-trusted environments. These algorithms require that the majority of the network be trustworthy and may suffer from message overhead as the network size increases [2].

Another issue is the cost (time and money) that healthcare providers are willing to accept to use a completely new paradigm in EHR management, as the industry is typically unwilling to invest money in supercomputers and in adopting new systems [8]. On top of all this, the European Union's *General Data Protection Regulations* (GPDR) of 2018 complicates the scenario.

A Blockchain ledger is permissioned or permissionless [14]. Permissionless means public: anyone can participate and the network is transparent. Permissioned means private: all participants are known and the network has restricted access. As healthcare deals with personal data, we need a permissioned one. Permissioned blockchains have desirable features for implementing modern EHRs [12]:

- **Decentralised**: Peer-to-peer (i.e., non-intermediated, P2P) architecture.
- **Smart Contracts**: A smart contract outlines the terms of relationships between peers with cryptographic code.
- **Immutable audit trail**: Trackable and timestamped patient-generated data.
- **Data provenance**: Evidenced source of medical research data.
- **Security/anonymity**: Patients control consent statements of their anonymously stored data, while confident that the network holds them securely.

There are various opportunities for blockchain-based applications in healthcare. Concretely, we can apply it to: (1) Longitudinal EHRs based on data from distant sources and providers, (2) automated health claims adjudications, (3) automatic providers interoperability, (4) secure online patient access to global EHRs, (5) farmaceutical supply chain managements [18], (6) tractability against counterfeit drugs industry, and (7) permissioned clinical trials and research database [13].

In this paper, we establish the foundations and develop a prototype of a low-cost, energy efficient, secure, interconnected, law compliant, and trustworthy EHRs sharing platform that empowers users. Technically speaking, ClinicAppChain[1] is a prototype built over Hyperledger Fabric and Composer Blockchain Framework using standard web-technologies, allowing conventional devices (e.g., computers, laptops, smartphones) to access ClinicAppChain applications and services, and acting as nodes of the ledger.Throughout this experience, we have identified and attempted to answer the following research questions (RQs):

---

[1] https://www.clinicappchain.com.

- **RQ1**: What are the foundations of Blockchain for healthcare?
- **RQ2**: How can media (e.g., images, videos. . .) be efficiently stored and inter-connected with a Blockchain ledger considering the GPDR'18?
- **RQ3**: Do we have the technology to build a sustainable and low-cost Blockchain solution for EHRs management in healthcare?

This paper presents the following structure. In Sect. 2 we review the state-of-the-art of current Blockchains for healthcare (RQ1). Then, we cope with data and media storage, legal aspects, compatibility (RQ2). In Sect. 3 we propose a model of a ledger for EHRs (RQ2), develop the prototype ClinicAppChain, and evaluate its energy footprint and costs (RQ3). In Sect. 4 we summarise the main conclusions and future work.

2   Related Work

Professional Blockchain projects are often built upon solid Blockchain frame-works [10], where the most mature in 2019, ordered by popularity are[2]:

1. **Ethereum** [19]: Widely known by its crypto-currency name (ETH) due to the popularity among the investors. It currently has the largest number of use-cases available. It is a permissionless P2P PoW (i.e., mined) based platform, which, like Bitcoin, is slow, energy-greedy, and needs high computation power, but is now switching to PoS, making it fit for private blockchains.
2. **Hyperledger Fabric** [3]: Hyperledger is an industrial grade framework with 185+ supporting companies of various industries, such as finance, supply chain management, Internet of Things, and manufacturing (e.g., The Linux Foundation, IBM, Intel). Fabric is its permissioned Business-to-Business (B2B) Blockchain Module, where peers can be grouped into organisations, and several concurrent ledgers can co-exist. It has no token or crypto-currency, and requires the majority of peers to consent to validate transactions.
3. **R3 Corda** [4]: Corda is a framework for the financial industry maintained by the company R3. It is a strict permissioned ledger, where just the peers involved in a transaction must validate it, clearly suited for banking and trading. Actual usage examples are HSBC and ING international banks.

Consequently, we present the most notable developments of blockchain healthcare ordered by advancements (RQs 1 & 2). **MedShare** [20] is a data sharing model between cloud service providers using Blockchain, which uses a private implementation of smart-contracts. In other words, the ledger contains permits alongside links to cloud data (e.g., EHRs), which is stored in databases or drives in their respective centres (e.g., hospitals, clinics, personal drives). This design is called *off-chain* [16,18], and it is, basically, a ledger that contains URLs to resources. **MediChain** [17] is an enterprise prototype of MedShare model. **MedRec** [6] is a proof-of-concept prototype of a Blockchain for off-chain EHRs, which uses Ethereum smart contracts. **BlocHIE** [9] is a dual loosely-coupled

---

[2] Survey: dzone.com/articles/best-3-enterprise-blockchain-platforms-for-rapid-p.

Blockchain for healthcare data exchange, having one ledger for EHRs hosted off-chain, and another for *in-chain* personal data. Lastly, **MedicalChain** [1] and **Proof.work**[3] are enterprise level off-chain Hyperledger Blockchain ecosystems with their own crypto-currencies (i.e., MED and PROOF) that aim to connect insurance, healthcare providers, and research institutions for certain fees. In contrast, our solution strategically combines on-chain and off-chain storage and uses no cryptocurrency, aiming for robustness and scalability.

Off-chain design is popular because it facilitates the adoption of a new system (interoperability vs a new standard) and due to the European GPDR'18. The most important point of this law in Blockchain systems is *the right to be forgotten*. A system must accept a citizen's request for complete data removal. However, Blockchain is a persistent system, so the off-chain solution leads to broken links. The parties storing the data (e.g., clinics and hospitals), are the ones to answer to removal requests. However, the security of a system is as weak as its weakest component. A traditional centralise database is a security hole, as just traditional security measures can be applied besides Blockchain. A breach on a single database could affect lots of EHRs, but on the other hand, media (e.g., images, videos) are expensive to store on-chain. We address this problem and propose a more balanced approach by keeping everything on-chain, except for large media files. As just media is off-chain, no written information (e.g., diagnostics) is compromised.

Every ledger solution and framework has been first developed and compiled for x86 architecture due to its market-share and applications compatibility. However, the most used in production are ported to ARM architecture except Hyperledger (community efforts are carried-out unsuccessfully[4]). Nonetheless, we are discussing computational nodes. Applications (i.e., service interfaces) can be developed in any programming language with front-end capabilities, and run on any compatible device (e.g., web services in a computer, smartphone, smart-tv...). Material and electricity costs depend on the computational requirements of the nodes running the consensus algorithm (e.g., miners in Bitcoin require extremelly powerful nodes). Generalising, when crypto-currencies are involved, as they are part of the profitability of the specific service, mining is needed to generate those tokens. Even if Bitcoin is the most profitable mined crypto-currency to date, expensive hardware and electricity costs exceed the currently rewarded value [15]. Instead of crypto-currencies, Hyperledger Fabric uses *Byzantine Fault Tolerance* algorithm, distributing consensus in 3 types of peers: Endorsement (i.e., users credentials), Validation (i.e., validate transactions), and Ordering (i.e., to sort validated transactions inside blocks).

---

[3] https://proof.work.

[4] Up to date, this was the only working effort that stopped to be compatible after Fabric 1.0 (2 years ago) https://github.com/Cleanshooter/hyperledger-pi-composer.

3   ClinicAppChain

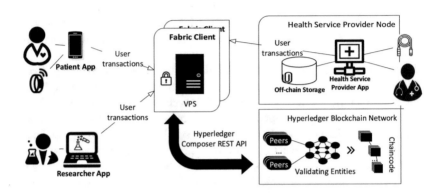

Fig. 1. ClinicAppChain - high-level architecture.

ClinicAppChain is our low-cost and energy-efficient prototype for an EHR management system. Healthcare providers work under rigorous privacy requirements for sensitive data, so permissionless frameworks are not an option. We have chosen Hyperledger Fabric (Fig. 1) version 1.4 because it is a fast (10k trans/sec.) and lightweight (no mining nor crypto-currencies) open source solution for permissioned Blockchains [7]. Moreover, it has a strong industrial and community support. The most important aspect of a Hyperledger implementation is the Business Network Definition (BND), which contains the data model definitions for all participants, assets, transaction logic, and access rules. The data model and access control rules are coded in *Hyperledger Composer Modelling Language*, and the transaction logic in JavaScript. Hyperledger Composer allows us to create a RESTful API for our Hyperledger system through the command line. Also, we can generate a basic user interface in Angular framework with *Yeoman*.

3.1   Blockchain Model

Patients, physicians, researchers, health insurers, pharmaceutical providers, governmental entities, are all actors (i.e., participants in a Blockchain jargon) that can potentially benefit from our EHR management system. This prototype currently focuses on just the first three actors. In Fig. 2 we illustrate the key use case scenarios in which the three actors interact with each other and the system (RQ1).

A **Patient** can request the following transactions: (i) edit *Personal Data*, which includes identification details as lifestyle habits, blood type, weight, height, allergies, and an emergency contact among others; (ii) consult the public profile of any Physician; (iii) add/remove a wearable device from its account; (iv)

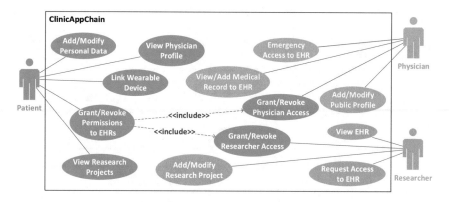

Fig. 2. ClinicAppChain use-case diagram. We use blue, green and grey to depict the use cases in which the Patient, Physician, and Researcher actor respectively, interact with each other and with the ClinicAppChain System

visualise research projects details; and, (v) grant and revoke permissions of his EHR to registered users (i.e., Physicians and Researchers). Permits can be granted/revoked collectively (to all Physicians/Researchers) or individually (to a specific Physician/Researcher).

A **Physician** can access the application to: (i) edit her *Public Profile*, including identification details, medical specialities, medical license and picture; (ii) access the EHR of a Patient in case of an accident, given that the patient is not conscious to explicitly give the consent and that the Physician is registered an emergency personnel; (iii) consult and add medical records to the EHRs of those patients that have permitted to do so; and, both patient and physician can list the spoken languages in their Personal Data and Public Profile respectively.

A **Researcher** can, (i) request the access to the EHR of any Patient; (ii) view the EHRs of Patients that have granted him access to their EHRs; and, (iii) add new research projects or modify the details of existing ones.

The central *asset* in ClinicAppChain is the EHR. Other assets are: Patient's Personal Data, Physician's Public Profile, and Research Projects. Each asset has one or more owners among the participants, and has a set of general access rules associated which limits who can do what transaction in the system. All participants need to authenticate to initiate a transaction. In Fig. 3 we illustrate the most representative transaction scenarios of ClinicAppChain with a sequence diagram showing the regular interaction between a Patient and a Physician. Each transaction is recorded in the Hyperledger's Blockchain.

Our greatest novelty is the mixed assets storage where we store text-based assets (e.g., diagnostics) on-chain, and media assets off-chain (e.g., MRI scan, X-Ray), unlike other prototypes discussed in the Related Work section (RQ2). This assures that, in case of a breach on any hospital, it only affects diagnostics's support data. In case of a removal request, the transaction history cannot be erased, but current data-set would be blank, and old data would be totally inaccessible, as with lost Bitcoins worth 1000 million US dollars to date [5].

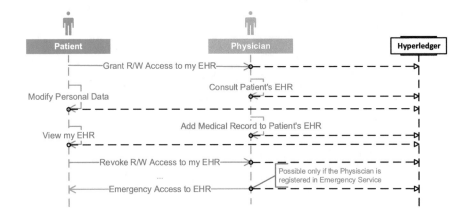

Fig. 3. ClinicAppChain general sequence diagram: Shows a typical course of interaction between a Patient (blue) and a Physician (green) in a healthcare ledger of Hyperledger

3.2   Evaluation

We have tested ClinicAppChain **peers** on three low-cost platforms: two virtual private servers, each featuring an AMD Opteron Processor 6176 with 2 cores @2.30 GHz and 4GB RAM, running Ubuntu Server 18.04 LTS x86_64, and a mini-PC worth 200\$, featuring an ultra low-power Intel i3-6100U @2.3 GHz Dual Core with Hyperthreading and 8GB RAM, running LUbuntu 18.04 LTS x86_64. For testing purposes, every node hosts the three peers at the same time, so any of them can perform as an endorsement, validation or ordering peer. They all match an average over the theoretical minimum value of **3.5k trans/sec**, so, a cheap low-power computer is capable enough for single peers. Regarding energy-footprint, we physically monitored the energy consumption of the mini-PC with *Watts Up? Pro*[5], resulting in an average energy consumption of **7.7 W** (RQ3). Additionally, we have tested the **client** (Angular front-end) in a low-power Raspberry Pi 3B+, featuring four cores Cortex-A53 (ARMv8 $times$ 64) and 1GB of RAM and representing the low-end segment of smartphones. On this 'wimpy' architecture, ClinicAppChain exhibits a negligible delay due to the wifi connection, with peaks of energy consumption of **3 W**, similar to requesting a blog web-page (RQ3). As mentioned, ARM architecture is not yet supported as a peer node in a Hyperledger Fabric Network.

4   Conclusions and Future Work

We identify the building blocks of a Blockchain solution for healthcare (Figs. 1 and 2) and develop ClinicAppChain, a permissioned Hyperledger Fabric prototype for EHR management (RQ1). We propose a novel mixed on-chain/off-chain ledger data partition, complying the EU GPDR'18 (RQ2). Using Fabric

---

[5] https://www.vernier.com/products/sensors/wu-pro/.

framework, we have obtained a low-cost solution —nodes can run on cheap and small hardware (e.g., mini-PC, Raspberry Pi), and energy efficient—$\sim$7.7 Watts per peer node, and $\sim$3 Watts per client node (RQ3). As we have developed the client front-end using web-technologies as HTML/CSS/JavaScript/Angular, any device with a web browser can log-in and make requests to the ledger (cross-platform). Our planned future work comprises: (1) allowing wearables and sensors (e.g., endoscope, heart-rate sensor) to automatically add patients' information to the EHR, where the Raspberry facilitates it as it has specific PINs, Bluetooth and USB ports, (2) expanding the participant's roles, transactions and EHR complexity, and (3) evaluating the usability of ClinicAppChain for a dentistry scenario to see how it can benefit in everyday clinical practice.

Acknowledgements. Work by Munoz and Fuentes is supported by the projects MAGIC P12-TIC1814 and HADAS TIN2015-64841-R (co-financed by FEDER funds). Work by Constantinescu and Asenjo is supported by the project TIN2016-80920-R, funded by the Spanish Government. ClinicAppChain has been supported by IMFAHE Foundation-Nodal Award of 2018. We are particularly grateful to other members of the awarded team that have guided us in the sanitary side and legal aspects: Esther Herrera, Laura Timanfaya, Patricia Rodríguez and Myriam Martínez.

References

1. Albeyatti, A.: White paper: medicalchain. MedicalChain self-publication (2018)
2. Andoni, M., Robu, V., Flynn, D., Abram, S., Geach, D., Jenkins, D., McCallum, P., Peacock, A.: Blockchain technology in the energy sector: a systematic review of challenges and opportunities. Renew. Sustain. Energy Rev. **100**, 143–174 (2019)
3. Androulaki, E., Barger, A., Bortnikov, V., Cachin, C., Christidis, K., De Caro, A., Enyeart, D., Ferris, C., Laventman, G., Manevich, Y., et al.: Hyperledger fabric: a distributed operating system for permissioned blockchains. In: Proceedings of the Thirteenth EuroSys Conference, p. 30. ACM (2018)
4. Brown, R.G., Carlyle, J., Grigg, I., Hearn, M.: Corda: An Introduction. CEV (2016)
5. Decker, C., Wattenhofer, R.: Bitcoin transaction malleability and MtGox. In: European Symposium on Research in Computer Security, pp. 313–326. Springer (2014)
6. Ekblaw, A., Azaria, A., Halamka, J.D., Lippman, A.: A case study for blockchain in healthcare: "medrec" prototype for electronic health records and medical research data. In: Proceedings of IEEE Open & Big Data Conference, vol. 13, p. 13 (2016)
7. Hintzman, Z.: Comparing blockchain implementations. In: Cable-Tec Expo (2017)
8. Iansiti, M., Lakhani, K.R.: The truth about blockchain. Harvard Bus. **95**(1), 118–127 (2017)
9. Jiang, S., Cao, J., Wu, H., Yang, Y., Ma, M., He, J.: BlocHIE: a blockchain-based platform for healthcare information exchange. In: 2018 IEEE International Conference on Smart Computing (SMARTCOMP), pp. 49–56. IEEE (2018)
10. Kolvart, M., Poola, M., Rull, A.: Smart contracts. In: The Future of Law and ETechnologies, pp. 133–147. Springer (2016)
11. Kombe, C., Ally, M., Sam, A.: A review on healthcare information systems and consensus protocols in blockchain technology. Int. J. Adv. Technol. Eng. Explor. **5**(49), 473–483 (2018)

12. Kuo, T.T., Kim, H.E., Ohno-Machado, L.: Blockchain distributed ledger technologies for biomedical and health care applications. J. Am. Med. Inf. Assoc. **24**(6), 1211–1220 (2017)
13. Mettler, M.: Blockchain technology in healthcare: the revolution starts here. In: 2016 IEEE 18th International Conference on e-Health Networking, Applications and Services (Healthcom), pp. 1–3. IEEE (2016)
14. Michael, J., Cohn, A., Butcher, J.R.: Blockchain technology. J. **1**, 7 (2018)
15. O'Dwyer, K.J., Malone, D.: Bitcoin mining and its energy footprint. In: Proceedings of the 25th IET Irish Signals & Systems Conference, IET (2014)
16. Poon, J., Dryja, T.: The bitcoin lightning network: scalable off-chain instant payments. Lightning Network self-publication (2016)
17. Rouhani, S., Humphery, D.G., Butterworth, L., Deters, R., Simmons, A.D.: Medichain: a secure decentralized medical data asset management system. J., 1533–1538 (2018)
18. Stagnaro, C.: Innovative blockchain uses in health care. Freed Associates (2017)
19. Wood, G.: Ethereum: a secure decentralised generalised transaction ledger. Ethereum Proj. Yellow Pap. **151**, 1–32 (2014)
20. Xia, Q., Sifah, E.B., Asamoah, K.O., Gao, J., Du, X., Guizani, M.: Medshare: trustless medical data sharing among cloud providers b. IEEE Access **5**, 14757–14767 (2017)

Smart Contracts are More than Objects: Pro-activeness on the Blockchain

Giovanni Ciatto[1], Alfredo Maffi[1], Stefano Mariani[2], and Andrea Omicini[1(✉)]

[1] Alma Mater Studiorum–Università di Bologna, Bologna, Italy
{giovanni.ciatto,andrea.omicini}@unibo.it, alfredo.maffi@studio.unibo.it
[2] Università di Modena e Reggio Emilia, Modena, Italy
stefano.mariani@unimore.it

Abstract. In this paper we focus on the expressiveness of smart contracts (SC) and its role in blockchain technologies (BCT), by presenting Tenderfone, a prototypical blockchain platform providing SC as pro-active, time-aware, and asynchronous entities.

Keywords: Smart contracts · Pro-activeness · Asynchronous interaction

1 Introduction

Smart contracts (SC) were originally conceived to *replace* real-world contracts by automatising the process of checking compliance to the terms of the agreement reified by the contract [10]. In spite of the heterogeneity of the many SC implementations available – such as Ethereum [11] and HyperLedger Fabric [1] – all of them share a common trait: SC are always interpreted in Object-Oriented (OO) terms [7] as *reactive* objects encapsulating a state of affairs and the behaviour meant to check compliance to the contract "application logic". More precisely, SC typically behave as (replicated) *distributed objects* interacting *synchnronously*.

In this paper we argue that such a design choice suffers of two main flaws: *(i)* conceptually, it is not fully in line with the original intended purpose of SC: in fact, real-world contracts are often *pro-actively* monitored and enforced by third party authorities, which take appropriate action in case of violation; *(ii)* technically, it limits the practical expressiveness of SC: for instance, mainstream BCT does not support SC executing *time-related activities* such as scheduling periodic payments or delaying actions in time [5]—as is the case of real-world scenarios such as payment of household bills, opening/closing of public auctions, contracting of public services in a submission window, etc.

Accordingly, we propose a novel notion of *pro-active, time-aware,* and *asynchronous* SC: pro-activeness is achieved through encapsulation of control flow, and is preserved by switching to asynchronous interaction (message passing)

© Springer Nature Switzerland AG 2020
J. Prieto et al. (Eds.): BLOCKCHAIN 2019, AISC 1010, pp. 45–53, 2020.
https://doi.org/10.1007/978-3-030-23813-1_6

instead of method/procedure call (which is synchronous), while reactiveness to time is achieved through dedicated constructs enabling time-awareness and scheduling of delayed actions. As a proof-of-concept, we present Tenderfone [9], as a prototype BCT platform supporting the new notion of—SC along with a new language for programming SC built on top of Tendermint [8].

2   Features of Pro-active Smart Contracts

Let us consider the case of a public auction to be handled by a smart contract. The public administration opens the submission window for contracting of public services at a given time, then accepts incoming bids, to finally close the auction when the deadline expires—*without any external intervention needed*. Here *time* plays a central role, as well as the capability of the public administration to act *pro-actively* so to close the auction when due: upon creation, the SC starts its own flow of control to concurrently wait for the deadline while accepting incoming bids/withdrawals. Currently, neither *pro-activeness* nor *time-reactiveness* are supported by mainstream SC implementations of SC—hence the use case is *out of the expressive reach* of available BCT.

Pro-activeness amounts to the capability of making something happen, instead of merely reacting to events. For computational entities, a requirement for pro-activeness is *encapsulation of control flow*: so, proactive SC must have their own control flow, so as to compute independently of the external stimuli, and without locking the user or another SC in the wait for completion. Another related requirement is *asynchronous interaction*, since synchrony would amount to an implicit delegation of control flow. Indeed, pro-activeness is connected to *time-awareness*: instead of acting in response to an external stimuli, an action is performed at a given point in time. Also asynchrony is required, since any synchronous interaction may hinder proactive behaviour and reactiveness to time.

Summing up, to increase the expressiveness of SC towards proactiveness, three strictly intertwined ingredients are required: *encapsulation of control, time-awareness, asynchronous interaction*.

3   The Tenderfone Language

To inject pro-activeness, reactiveness to time, and asynchronous communication in SC we define a new language for programming SC—namely, the Tenderfone language, a declarative language based on Prolog. Accordingly, data encapsulated as part of the SC state are expressed as Prolog terms, which constitute its Knowledge Base (KB) as split in two: a *static* partition represents the program to be executed; the *dynamic* one represents its (encapsulated) mutable state. Then, a Tenderfone SC can be interpreted as a logic reasoner proving goals either pro-actively pursued or delegated by end users or other SC: in fact, the associated Tenderfone Interpreter (Sect. 4) is essentially a logic-based reasoning engine exploiting tuProlog [6], following the approach outlined in [4].

3.1 Declarative Smart Contracts

A Tenderfone SC is a Prolog theory exposing a number of *entry points* (Prolog rules) to be implemented by developers and exploiting a number of *built-in* functions ready to use. Entry points come in the form "init(+Args):- Body" or "receive(+Msg, @Sender):- Body", where init is triggered (just once) right after *deployment* and endows the SC with it own *flow of control*; receive is executed whenever *(a)* a user publishes an invocation transaction or *(b)* the SC is the recipient of a message from a SC (there including itself)—as depicted in Table 1. Messages can be sent through the send(+Msg, +Recipient) built-in, with an *asynchronous* semantics: receive is *eventually* triggered on the recipient SC if an only if execution of the sender's entry point (where send is called) terminates successfully. This implies that the recipient entry point is always executed *after* the sender's one—even if they coincide.

Table 1. Tenderfone language at a glance (*message* is a new kind of transaction defined in Tenderfone)

Entry point	Executor	Trigger	Transaction	Initiator
init(A)	SC with $KB = T$	API deploy(T, A)	Deployment	User
receive(M, S)	SC R	API invoke(M, R)	Invocation	User S
		built-in send(M, R)	*message*	SC S
receive(M, R)	SC R	built-in when(T, M)	*message*	SC R
		built-in delay(D, M)	*message*	SC R
		built-in periodically(P, M)	*message*	SC R

Reactiveness to time is based on built-in predicates: *(a)* now(-Timestamp) lets SC fetch current time; *(b)* when(@T, +Msg) triggers a receive of the calling SC at time T; *(c)* delay(@DT, +Msg) equals now(T), when(T+DT, Msg); *(d)* periodically(@P, +Msg) triggers a calling SC receive every P time units.

Figure 1 shows an example of a Tenderfone SC that delays auction closing until global time reaches a given instant in the future. It keeps collecting proposals, but only while the auction is open, refusing them otherwise. This represents a meaningful combination of asynchronous message passing and time-*reactiveness* aimed at handling an auction with no human-in-the-loop. Furthermore, such a behaviour is *pro-active* since the SC itself schedules its own behaviour by means of the delay/2 primitive, within init.

```
init(Period) :- % Period known by construction
  set_data(auction_status, open), % ensure auction status
  delay(close, Period). % schedule acution closing

receive(proposal(Bid), S) :-
  get_data(auction_status, open), % ensure auction is open
  % handle proposal

receive(close, Me) :- % when it is time to close
  self(Me), % only this SC can close the auction
  set_data(auction_status, closed). % update auction status
```

Fig. 1. The Tenderfone smart contract for the public auction use case

3.2   Requirements for Tenderfone BCT

In order to support the Tenderfone language, the underlying BCT should satisfy some specific requirements. BCT are essentially distributed virtual machines executing SC and keeping track of the state of all them, as deployed by end users during system operation. The state of a particular SC may then evolve as a consequence of end users *triggering* computations by means of *transactions*.

At this level of abstraction, a few assumptions are satisfied by all 2nd and 3rd generation BCT: (**A1**) transactions are *always* generated by end users; (**A2**) SC are triggered by transactions only; (**A3**) SC are triggered by end users only—as a consequence of **A1** and **A2**; (**A4**) once triggered, SC can observe time, and use it as an input for their computation. To support Tenderfone SC language, we need to replace **A1** with (**R1**) transactions may be generated by both end users *and SC*; which, combined with assumption **A2** leads to (**R3**) SC can be triggered by both end users *and SC*.

The following section explains how satisfying **R1** and **R3** while preserving **A2** and **A4** is *sufficient* to support the Tenderfone language.

4   The Tenderfone Platform

In Tenderfone, we model the blockchain as a distributed system made of a set of *users* $\mathcal{U}$ interacting with a set of *validators* $\mathcal{V}$. Validators in $\mathcal{V}$ enact a consensus protocol – among the many available in the literature [2,3] – meant to let them all agree on the current system state, which evolves at discrete time steps so that the system as a whole perceives the sequence $S_0, \ldots, S_i \in \mathcal{S}$ of *consistent* states. Each transaction $tx \in \mathcal{T}$ may produce a state change from some S to an *intermediate* state S' through transition function $\gamma : \mathcal{S} \times \mathcal{T} \to \mathcal{S}$ encapsulating the semantics of transactions execution. Because of distribution, the many $\mathcal{V}$ may perceive different *orderings* for $tx_1, tx_2, \ldots$ possibly sent by users in the same time window. Each $V \in \mathcal{V}$ drops *invalid* transactions according to an application-level policy embedded in the *validation* relation $\nu \subseteq \mathcal{S} \times \mathcal{T}$—we write $\nu(S, tx)$ for each tx valid in S. In order for all the validators to agree on the most recent state S_i, they must timestamp and register on the ledger (within *blocks*) the sequence of transaction leading from S_0 to S_i through γ, for all time steps i. Hence, Tenderfone *block-chain* consists of a sequence $B_0, \ldots, B_i \in \mathcal{B}$ of blocks, and function $\Gamma : \mathcal{S} \times \mathcal{B} \rightarrow \mathcal{S}$ maps B_i upon $S_i = \Gamma(S_{i-1}, B_i)$, which computes S_i by recursively applying all transactions through γ.

Each validator V continuously performs the following activities, concurrently: *(a)* accept transactions, validating each one, dropping the invalid ones, and temporarily storing the valid ones until they are *eventually* included into some block; *(b)* engage in consensus with the other validators to order the valid transactions received so far, and, therefore, agree on the content of next block B_{i+1}; *(c)* execute transactions in B_i to update the system state. All these activities contribute to support life-cycle management and execution of Tenderfone smart contracts.

Accordingly, $S_i \in \mathcal{S}$ consists of the set $\mathbf{K}_i = \{K_{i,1}, \ldots, K_{i,N_i}\}$ of the SC knowledge bases, where N_i is the number of SC deployed until time step i. In this setting, through function γ, a transaction $tx \in \mathcal{T}$ may trigger a computation on SC $K \in \mathbf{K}_i$ or deploy a new SC K'—given its KB and initialisation arguments.

In the end, validators have two duties: running consensus and executing SC code. Thus, a validator $V \in \mathcal{V}$ can be designed as two processes: C_V (the *consensus engine* in charge of collecting and ordering transactions) and I_V (the *SC interpreter* responsible for their validation and execution)—hence $V = \langle C_V, I_V \rangle$.

4.1   The Consensus Engine

The consensus engine C_V interacts with the interpreter I_V as the client of a request-response protocol for which I_V is the server. The steps are as follows.

Transactions Validation. Whenever a transaction tx is sent by a user U to some C_V, the latter propagates it to all $C_{V'} : V' \in \mathcal{V} - \{V\}$ by means of an arbitrary *gossiping* algorithm. Upon reception, tx is forwarded to I_V for validation via a $\mathtt{checkTx}(tx)$ message. I_V then validates it against the latest committed state S_i and answers with a $\mathtt{checkTxResponse}(r)$ where $r = \mathtt{true} \iff \nu(S_i, tx)$ holds. In case $r = \mathtt{true}$, C_V includes tx into its local storage area (its *mempool* M_V) to be later included into some committed block. In case $r = \mathtt{false}$, tx is dropped.

Blocks Management. When agreement on the content of block B_{i+1} is needed, the J_{max} oldest transactions are popped from M_V and included into C_V's candidate block B_{i+1}^V. Until the number of transactions in M_V is $< J_{max}$, a candidate block is anyway produced once every ΔT_{max} time units. As soon as block B_i is jointly produced, each C_V delegates execution of its transactions to I_V. To do so, the following request-response protocol takes place:

1. C_V sends a message $\mathtt{beginBlock}(i, T_i)$ to I_V, telling that the ith block has been agreed upon, and that its timestamp is T_i—hence, the global (logical) time reached T_i
2. I_V sends back a message $\mathtt{beginBlockResponse}(\mathbf{tx'})$, informing C_V that a (possibly empty) list of transaction $\mathbf{tx'}$ have been generated *because* the global time reached T_i

3. for each $tx_{i,j} \in B_i$, C_V sends a message deliverTx($tx_{i,j}$) to I_V to request execution of $tx_{i,j}$. Then, I_V computes and stores $S_{i,j} = \gamma(S_{i,j-1}, tx_{i,j})$, where $S_{i,0} \equiv S_i$

4. C_V sends a message endBlock(i) to I_V, telling that all transactions in $\mathbf{tx}_i$ have been processed thus it can now perform application-level housekeeping operations, if any

5. C_V sends a message commitBlock(i) to I_V, asking I_V to compute $hash(S_i)$

6. I_V sends back a message commitBlockResponse($s_i, \mathbf{tx}''$) informing C_V that a (possibly empty) list of transaction $\mathbf{tx}''$ have been generated as a consequence of block B_i being committed

It is worth to emphasise that transactions generated by I_V in steps 2 and 6 are treated by C_V as regular transactions, as if they were received from users, hence are eventually propagated to all other $C_{V'} : V' \in \mathcal{V} - \{V\}$, validated by I_V, and executed by I_V. As step 2 transactions are generated from when, delay, and periodically built-ins and step 6 ones stem from send, those steps represent the very essence of the mechanism enabling asynchronous message passing and time-reactiveness for Tenderfone SC—hence what makes Tenderfone satisfy requirement **R1** from Subsect. 3.2.

Finally, let us highlight that each transaction tx possibly generated by some I_V in steps 2 and 6 is actually registered many times on the ledger, since the many $I_V : V \in \mathcal{V}$ are all replicating the same protocol. This is an intended behaviour, mean to prevent corrupt validators from forging fake transactions (*spurious* transactions), as described in Subsect. 4.2.

Tenderfone vs. Tendermint. The consensus engine described above is a generalisation of (and almost a precise mapping to) Tendermint "core" abstraction. In Tendermint, in fact, validators are called "nodes", consensus engines are called "cores", whereas smart contract interpreters are called "apps" [8]. There, cores and apps interact by means of the ABCI interface[1]—that is, a HTTP interface were cores act as clients and apps act as servers. End users are expected to interact with Tendermint cores by means of a HTTP interface, too.

Steps 2 and 6 are the only notable differences between Tenderfone and Tendermint. In fact, the latter does not allow for transactions to be generated by apps, neither are cores able to handle transactions returned by apps. To overcome such limitation of Tendermint, Tenderfone – available for public scrutiny at [9] – let apps behave as ordinary end users.

4.2 The Tenderfone Interpreter

The Tenderfone interpreter I_V is the software component in charge of locally executing SC in a transaction-driven way. To do so, it maintains the data structures constituting the system state $S_i \in \mathcal{S}$ discussed above, and updates them depending on the information contained in each message sent by C_V.

---

[1] https://tendermint.com/docs/app-dev/app-architecture.html.

More precisely, an *intermediate* system state – that is, the system after i blocks have been committed, and j more transactions have been processed since then – is defined as a 5-uple $S_{i,j} = \langle T_i, \mathbf{K}_{i,j}, \mathbf{Q}_{i,j}, \mathbf{M}_{i,j}, \mathbf{\Theta}_{i,j} \rangle$ such that:

- $T_i \in \mathbb{R}_{>0}$ is the timestamp of block B_i
- $\mathbf{K}_{i,j} \subset \mathcal{K}$ is the set of the knowledge bases corresponding to the SC *currently* deployed
- $\mathbf{Q}_{i,j} \in \mathcal{Q}^*$ is the queue of scheduled tasks which are *currently* waiting to be executed
- $\mathbf{M}_{i,j} \in \mathcal{M}^*$ is the queue of outgoing messages which are *currently* waiting to be delivered
- $\mathbf{\Theta}_{i,j} : Supp(\mathbf{\Theta}_{i,j}) \subseteq \mathcal{T}$ is a multiset tracking the many copies of the generated transactions

Assuming block B_i contains J_i transactions, we say that $S_i \equiv S_{i,J_i} = \Gamma(S_{i-1}, B_i)$ is the state reached by sequentially applying all transactions in B_i to $S_{i-1} \equiv S_{i,0}$.

The behaviour of I_V is then informally defined as follows, in terms of how state $S_{i,0}$ is updated as a consequence of I_V handling C_V's requests.

Schedule Computations. Whenever I_V receives a message `beginBlock`$(i+1, T_{i+1})$ by C_V, it selects the scheduled tasks in $\mathbf{Q}_{i,J_i}$ which are ready to be executed according to the new timestamp T_{i+1} and *generates* a list of transactions $\mathbf{tx}'$. It then computes a novel state $S_{i+1,0}$ which is equal to S_{i,J_i} except for T_i, and $\mathbf{Q}_{i,J_i}$ Finally, it sends back to C_V a message `beginBlockResponse`$(\mathbf{tx}')$ returning to C_V the transactions generated out of scheduled tasks. Notice that scheduled tasks may be added to $\mathbf{Q}_{i,J_i}$ during transaction execution by calling `when`, `delay`, or `periodically`.

Execute Transactions. Whenever I_V receives a message `deliverTx`$(tx_{i,j})$ by C_V, requesting the execution of transaction $tx_{i,j}$, the actual effect depends on the nature of transaction $tx_{i,j}$ and on *who* sent it. First of all, $\mathbf{\Theta}_{i,j}$ is computed as $\mathbf{\Theta}_{i,j-1} \cup \{tx_{i,j}\}$: this is necessary because, if $tx_{i,j}$ sender is a SC, then it is actually processed as described below *only if* the multiplicity of $tx_{i,j}$ in $\mathbf{\Theta}_{i,j}$ is equal to the number of validators $\mathcal{V}$—to prevent corrupted validators from forging spurious transactions. Then:

- in case of a *deployment* transaction, $tx_{i,j}$ carries the KB K of a novel SC along with the argument A for its initialisation. `init` is then invoked on K using A as argument and, if it terminates successfully, $\mathbf{K}_{i,j}$ is computed as $\mathbf{K}_{i,j-1} \cup \{K'\}$, being K' the new version of K containing the side effects possibly produced by `init` itself
- in case of an *invocation* transaction, $tx_{i,j}$ carries the identifier id_K of the SC whose KB is K, and a message M to be handled by id_K receives. A specific `receive` is then invoked on K based on M and id_K and, if it terminates successfully, $\mathbf{K}_{i,j}$ is computed as $(\mathbf{K}_{i,j-1} - \{K\}) \cup \{K'\}$, being K' the new version of K containing the side effects possibly produced by `receive` itself

In any case, execution of entry points may step through one or more send invocations. If so, $\mathbf{M}_{i,j}$ is computed by adding to $\mathbf{M}_{i,j-1}$ an outgoing message for each send. Similarly, $\mathbf{Q}_{i,j}$ is computed by adding to $\mathbf{Q}_{i,j-1}$ a new scheduled task for each when, delay, or periodically invocation.

Housekeeping. Whenever I_V receives a message endBlock(i) it computes $\Theta_{i,j}$ so that all transactions in $\Theta_{i,j-1}$ whose multiplicity equals the number of validators $\mathcal{V}$ are removed from $\Theta_{i,j-1}$.

Dispatch Messages. Whenever I_V receives a message commitBlock(i) by C_V, it *generates* a list of transactions $\mathbf{tx}''$ based on outgoing messages stored in $\mathbf{M}_{i,J_i}$. It then updates state S_{i,J_i} in such a way that $\mathbf{M}_{i,J_i}$ is emptied. Finally, it sends back to C_V a message commitBlockResponse($s_i, \mathbf{tx}''$) where $s_i = hash(S_{i,J_i})$, in order to return to C_V the transactions generated from outgoing messages.

5   Conclusion

The mainstream notion of smart contract (SC) comes with several limitations hindering its full potential. In this paper we start from the observation that object-orientation is a poor design choice for SC, preventing them from performing, e.g., time-related tasks – like waiting for a given moment in the future – or any other task where avoiding human-in-the-loop could be desirable—e.g., for auctions or public tenders. Accordingly, we show how the internal function of a blockchain technology (BCT) – and in particular the component aimed at executing SC – can be designed in order for SC to act pro-actively or in a time-aware way. To do so, we present Tenderfone [9], a BCT prototype supporting logic-based, pro-active smart contracts on top of Tendermint [8].

References

1. Androulaki, E., et al.: Hyperledger fabric: a distributed operating system for permissioned blockchains. In: 13th EuroSys Conference (EuroSys 2018). ACM, New York (2018). https://doi.org/10.1145/3190508.3190538
2. Aublin, P.L., Guerraoui, R., Knežević, N., Quéma, V., Vukolić, M.: The next 700 BFT protocols. ACM Trans. Comput. Syst. **32**(4), 1–45 (2015). https://doi.org/10.1145/2658994
3. Cachin, C., Vukolić, M.: Blockchain consensus protocols in the wild (keynote talk). In: 31st International Symposium on Distributed Computing (DISC 2017). LIPIcs, vol. 91, pp. 1:1–1:16. Dagstuhl, Germany (2017). https://doi.org/10.4230/LIPIcs.DISC.2017.1
4. Ciatto, G., Calegari, R., Mariani, S., Denti, E., Omicini, A.: From the blockchain to logic programming and back: research perspectives. In: Cossentino, M., Sabatucci, L., Seidita, V. (eds.) WOA 2018 – 19th Workshop From Objects to Agents, CEUR Workshop Proceedings, vol. 2215, pp. 69–74 (2018). http://ceur-ws.org/Vol-2215/paper_12.pdfs

5. Ciatto, G., Mariani, S., Omicini, A.: Blockchain for trustworthy coordination: a first study with Linda and Ethereum. In: 2018 IEEE/WIC/ACM International Conference on Web Intelligence (WI), pp. 696–703 (2018). https://doi.org/10.1109/WI.2018.000-9

6. Denti, E., Omicini, A., Ricci, A.: tuProlog: a light-weight prolog for internet applications and infrastructures. In: practical aspects of declarative languages. LNCS, vol. 1990, pp. 184–198. Springer (2001). https://doi.org/10.1007/3-540-45241-9_13

7. Governatori, G., Idelberger, F., Milosevic, Z., Riveret, R., Sartor, G., Xu, X.: On legal contracts, imperative and declarative smart contracts, and blockchain systems. Artif. Intell. Law **26**(4), 377–409 (2018). https://doi.org/10.1007/s10506-018-9223-3

8. Kwon, J.: Tendermint: Consensus without mining (2014). https://tendermint.com/static/docs/tendermint.pdf

9. Maffi, A.: Tenderfone GitLab Repository. https://gitlab.com/pika-lab/blockchain/tenderfone/tenderfone-sc

10. Szabo, N.: Smart contracts (1994). http://www.fon.hum.uva.nl/rob/Courses/InformationInSpeech/CDROM/Literature/LOTwinterschool2006/szabo.best.vwh.net/smart.contracts.html

11. Wood, G.: Ethereum: a secure decentralised generalised transaction ledger (2014). http://ethereum.github.io/yellowpaper/paper.pdf

Blockchain Based Informed Consent with Reputation Support

Hélder Ribeiro de Sousa[1(✉)] and António Pinto[1,2]

[1] CIICESI, ESTG, Politécnico do Porto, Porto, Portugal
helderribeirosousa@gmail.com
[2] CRACS and INESC TEC, Porto, Portugal
apinto@inesctec.pt

Abstract. Digital economy relies on global data exchange flows. On May 25th 2018 the GDPR came into force, representing a shift in data protection legislation by tightening data protection rules. This paper introduces an innovative solution that aims to diminish the burden resulting from new regulatory demands on all stakeholders. The presented solution allows the data controller to collect the consent, of a European citizen, in accordance to the GDPR and persist proof of said consent on public a blockchain. On the other hand, the data subject will be able to express his consent conveniently through his smartphone and evaluate the data controller's performance. The regulator's role was also contemplated, meaning that he can leverage certain system capabilities specifically designed to gauge the status of the relationships between data subjects and data controllers.

Keywords: GDPR · Informed consent · Reputation systems ·
Blockchain · Ethereum platform

1 Introduction

Today's digital economy relies heavily on global data exchange flows. This reality was built up by organizations operating worldwide that, in order to gain competitive advantage and simplify internal processes, went on to capture, process and store its users personal data. By doing so, they were obliged to comply with complex legislation that could vary greatly depending on jurisdiction. Sometimes those organizations also had to take in consideration the fact that some of that legislation was not clear [1]. Recent privacy related scandals such as Cambridge Analytica's, a company that from 2014 to 2018 inordinately leveraged the data of at least 87 million Facebook users to manipulate the outcome of public elections in both the United States of America and the United Kingdom [2], and

This work was partially financed by National Funds through the Portuguese funding agency, FCT - Fundação para a Ciência e a Tecnologia within project UID/EEA/50014/2019.

J. Prieto et al. (Eds.): BLOCKCHAIN 2019, AISC 1010, pp. 54–61, 2020.
https://doi.org/10.1007/978-3-030-23813-1_7

America's retail credit giant Equifax [3], that in September of 2017 saw financial data of 143 million citizens illegally accessed, are cited by the civil society as undeniable proof of personal data negligence and abuse perpetrated by the entities to whom we concede our data.

Such phenomena grabbed the European Union (EU) attention and led them to impose the General Data Protection Regulation (GDPR) [4] in an attempt to prevent such events. The GDPR represented a major shift in data protection by tightening data protection rules [5]. Any organization processing personal data of European citizens must comply with it. The informed consent as imposed by the GDPR is of particular importance. Consent must be given by the data subject in a clear affirmative act, establishing a specific and informed indication of agreement to the processing of his personal data. When the processing has multiple purposes, consent should be given for each and all of them. Furthermore, if the data processing is based on the data subject's consent, the controller must be able to present proof that the consent was given by the data subject. Infringing organizations may face fines of millions of Euros. Regulators will have full autonomy to sanction infringing organizations accordingly.

While organizations strive to adapt their operations to a post-GDPR reality, blockchain technology flourishes and presents itself as the *de facto* solution for distributed and decentralized systems that require an immutable ledger. Different industries rapidly identified prominent business opportunities that such technology arose [6]. Companies from different business areas such as banking, real-estate, insurance and internet of things joined efforts to promote the research and development of blockchain technology for multiple purposes. Their goal is to integrate blockchain technology on their systems in order to fulfill their own business' needs and meet regulatory demands [7].

In particular, blockchain technology appears as a form of mediation between multiple parties with conflicting interests. This is the case of the informed consent imposed by GDPR where data processors want to collect data, data subjects must be able to authorize and regulators must be able to audit. The proposed system leverages blockchain technology to provide a solution to collect and persist informed consents in compliance with the GDPR and to empower data subjects through a reputation system that will allow them to evaluate the data controllers' performance. Proof of all transactions occurring in the system will be persisted on a public blockchain. Data subjects will interact with the system to receive and act upon consent requests issued by data controllers, remove previously given consents or evaluate data controllers. Data regulators will be able to access interactions between data subjects and data controllers.

2   Blockchain and Ethereum Platform

Satoshi Nakamoto published Bitcoin, a decentralized electronic cash system [8] that solved the double spending problem of digital currencies while being a fully decentralized system. Bitcoin went on to capture the public's attention and became the most successful cryptocurrency to date [9]. Individuals with different academic and professional backgrounds explored investment opportunities made possible by the initial cryptocurrency wave. The industry as also shown a keen interest in blockchain technology.

Blockchain is the generic designation given to the algorithms and cryptographic schemes that assure the integrity and traceability of all transactions that take place on the Bitcoin network without resorting to a central authority, thus allowing it to be distributed and fully decentralized. The idea of chaining blocks of cryptographically related data, as used by Nakamoto, prevented double-spending situations in the Bitcoin network through the definition of a proof-of-work consensus protocol. Most of the cryptocurrencies that followed chose a similar path [10].

The industry demand for blockchain technology resulted in what is now known as "Blockchain 2.0". This new blockchain generation included support for smart contracts and new consensus protocols, such as proof-of-stake and proof-of-authority. Companies could then explore such features to embed their business logic into the blockchain itself and develop decentralized applications that leveraged the platform's unique capabilities [11]. The Ethereum platform is one of the "Blockchain 2.0" implementations, providing a high-level programming language for smart contract definition. The language is inspired by C and Javascript and is named Solidity. Is also provides an integrated development environment, a compiler, client software, test networks, and a network explorer. The community followed with many other critical tools, such as the web3 libraries and the *Metamask* browser extension.

3   Trust and Reputation Systems

Manifestations of trust are easy to recognize because we experience and rely on them every day. At the same time, trust is quite challenging to define because it manifests itself in many different forms. There are two kinds of trust which are know as reliability trust and decision trust. Reliability trust can be interpreted as the reliability of something or somebody, and can be defined by the trust an individual A places on the expectation that another individual, B, will perform a given action on which his well being depends. Decision trust is the extent to which one party is willing to depend on something or somebody in a given situation with a feeling of relative security, even though negative consequences may arise [12].

The concept of reputation is closely linked to that of trustworthiness, but it is evident that there are important differences. Trust is a personal and subjective phenomenon that is based on various factors and evidences, and some of these weight more than others. Personal experience may weight more than second hand opinions or reputation, but in the absence of personal experience, trust is usually based on recommendations from others. Reputation can then be considered as the collective measure of trustworthiness based on the recommendations or ratings from members in a community. An individual's subjective trust can be derived from a combination of recommendations and personal experience [12].

The advent of e-commerce and the massive worldwide adoption of platforms such as those from eBay and Amazon in the last decade of the XX century, triggered the need for online reputation systems. The recent global smartphone adoption and the ubiquity of Internet access in such devices led to a second wave of adoption and use of reputation systems [13]. The key purpose of such systems is to allow people to assign a reputation indicator to a third party's trustworthiness. This reputation indicator is confined to a specific domain and opinions are expressed by assigning values to the other party's behavior in a particular transaction or interaction [14]. The origin of reputation systems goes back to real life interactions, whenever people exchange opinions regarding services or other people. Professional recommendation letters are another example of such. In their most recent versions, reputation systems are provided by the platforms themselves and are seen as a key module by the community, aiming at the experience exchange between users and thus fostering trust. In some cases, the instigation for usage of reputation systems goes to such lengths that discounts, loyalty points or similar benefits are offered to users that rate their interactions with other community members [15].

The way a user shares its experience may differ from one platform to another. Some adopt a classification scale while others expect users to write small reviews, whereas some others combine both approaches. The most common experience representation setups are: binary, numerical, symbolical, textual or audiovisual. In binary input systems, the user expresses if he liked the experience or not. In numerical, the user inputs a level from one to five stars, for instance. In symbolical, the user leaves feedback by clicking on a symbol that represents his experience, such as a thumbs up icon or a like button. In textual, the user inputs a short review in the form of a text. In audiovisual, the user provides a video or audio record of his spoken experience. Such information is then compiled and presented on the corresponding member's profile page for everyone to see.

4   Proposed Solution

The proposed solution, named "Connsent", aims to manage all stages of informed consent processes in accordance to the GDPR while allowing data subjects to evaluate data controllers through an embedded reputation system. Its architecture is depicted on Fig. 1. Proofs of consent and reputation evaluations taking place in the system are persisted on a public blockchain. This approach allows the leveraging of native blockchain capabilities to assure the integrity of persisted consents and evaluations. Data collectors are expected to pay for the transaction fees.

The system was designed with three different user types in mind: data subjects, data controllers and regulators. Even though their roles in the system are different, their enrollment process is identical. Before any of them can use the system, they must choose an universally unique identifier (UUID), generate an Elliptic Curve Digital Signature Algorithm (ECDSA) key-pair with a 224 bit strength and perform a digital signature over the chosen UUID and the ECDSA key-pair's public key. That data must then be sent to system through its web API in order for it to be persisted on the used blockchain. This initial data will be used to validate future authorization proofs issued by its owner. Each user uses the generated ECDSA key-pair's private key to perform digital signatures. The key-pair is generated whenever the application is required to use it for the first time. A signature scheme was defined for the authorization of each type of operation. After all intervening users are enrolled, further interactions with the Connsent system will happen strictly for consent and evaluation purposes.

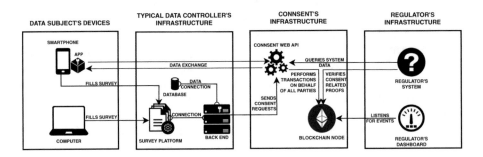

Fig. 1. Proposed solution's architecture

Interaction between a data subject and a data controller usually starts with the data subject entering some personal information into a data collection platform, such as a survey platform, ran by the data controller (as shown in Fig. 1). When this initial interaction is concluded, the data controller will send a consent request to the subject's device by invoking the system's web API with the necessary data. Such data includes his own identifier, the consent operation's UUID, the subject's UUID, the term's identifier, the regulator's identifier, the digest of the collected data, and a digital signature that attests his will to perform the request. If the invocation is successful, the system will then send a push notification to the subject's mobile device informing him that he received a new consent request.

The mobile application shown in Fig. 2 allows the data subject to act upon received consent requests, remove previously given evaluations, evaluate data controllers with whom he has interacted with, and view all data controllers' evaluations. Cryptographic proofs of those operations are generated within the mobile application using the user's private key and persisted on the blockchain by means of the Web API that supports the proposed solution.

Once the data subject consents to the processing of his data, he will be able evaluate the controller's performance through the reputation system. This design assures data controllers that they will only be evaluated by data subjects with whom they had legitimate interactions. The subject can leave a message describing his experience, plus a satisfaction indicator. The satisfaction indicator can be positive, neutral or negative. The controller's reputation score is calculated

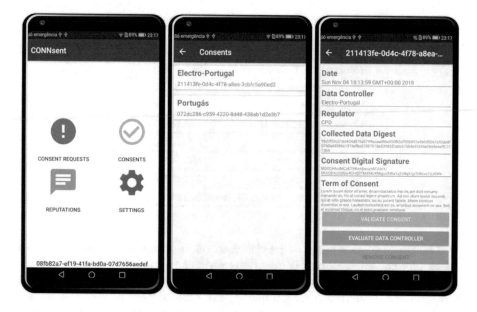

Fig. 2. Prototype mobile application

through the difference between the positive and the negative counts. The data subject can perform a new evaluation that will supersede the previous one. This information will be displayed on the controller's profile and available for everyone to see.

Regulators will be able to listen to contract account "NewConsent", "NewRemovedConsent", and "NewFeedbackEntry" events. "NewConsent" is used when a new consent under his supervision is given. "NewRemovedConsent" is used when a new consent under his supervision is removed. "NewFeedbackEntry" is used when a data subject evaluates a data controller regarding a consent under his supervision. These mechanisms allow the regulator to monitor operations in real time. The reputation system can be a tool to gauge the status of the relationships between data subjects and data controllers.

In order to persist the system's data on Ethereum blockchains a smart contract[1] was developed using Solidity. The smart contract enables the storage of data on the contract account's using custom data structures and specifies custom events to that will be emitted whenever system operations are stored on the contract account. Validation mechanisms that prevent any network member, other than the creator of the contract account, from performing transactions were set up. The smart contract was deployed on both Rinkeby (0x0fc33976d3d9b5e97a9d52c52a798ce8148aa9e2) and Ropsten (0x0fc33976d3d9 b5e97a9d52c52a798ce8148aa9e2) test networks. Transactions can be found on these contract accounts and their average cost is shown in Table 1.

---

[1] Available at https://pages.estg.ipp.pt/~apinto/connsent.sc.

Table 1. Average transaction cost by operation type

Operation type	Average gas cost
Controller creation	261,675.0
Regulator creation	260,930.0
Subject creation	303,408.4
Consent term creation	496,859.7
Consent	573,640.4
Consent removal	202,347.0
1st consent evaluation	321,848.0
Following consent evaluations	249,534.0

The usage of mappings to persist data on the contract account's storage allowed for the validation of transactions without requiring to traverse existing data. Such lead to transaction costs with minimum variation over time. This approach was preferred after comparing the cost of persisting the same one hundred transactions in a contract account originated by a smart contract that used arrays with another one that opted by mappings. The usage of mappings allowed transaction costs to remain virtually constant while the usage of arrays led to a linear cost increase (see Fig. 3).

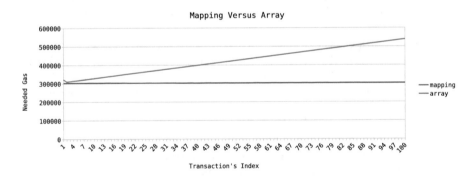

Fig. 3. Mapping vs array comparison

5 Conclusions

This paper introduced an innovative solution that aims to diminish the burden resulting from new privacy related regulatory demands on all stakeholders. It allows the data controller to collect a European citizen's consent in accordance to the GDPR and persist proof of said consent on the blockchain. On the other hand, the data subject will be able to express his consent conveniently through his smartphone and evaluate the data controller's performance. The regulator's role was also contemplated, meaning that he can leverage certain system capabilities specifically designed to gauge the status of the relationships between data subjects and data controllers. The adoption of the Ethereum Platform and the usage of specific smart contract definition features allowed for virtually constant transaction costs, thus enabling the forecast of the system's running costs.

References

1. Sater, S.: Blockchain and the European Union's General Data Protection Regulation: A Chance to Harmonize International Data Flows. SSRN (2017)
2. New York Times: Facebook says cambridge analytica may have gained 37m more users' data (2018). https://www.theguardian.com/technology/2018/apr/04/facebook-cambridge-analytica-user-data-latest-more-than-thought. Accessed 10 Apr 2018
3. CNN: Giant Equifax data breach: 143 million people could be affected. http://money.cnn.com/2017/09/07/technology/business/equifax-data-breach/index.html (2017). Accessed 31 Jan 2018
4. European Union: Regulation (EU) 2016/679 of the European Parliament and of the Council (2016). http://eur-lex.europa.eu/legal-content/EN/TXT/?uri=CELEX:32016R0679. Accessed 31 Jan 2018
5. Ibanez, L.-D., O'Hara, K., Simperl, E.: On blockchains and the general data protection regulation. Project report, July 2018
6. Mattila, J., et al.: The blockchain phenomenon–the disruptive potential of distributed consensus architectures. Technical report, The Research Institute of the Finnish Economy (2016)
7. Aste, T., Tasca, P., Di Matteo, T.: Blockchain technologies: the foreseeable impact on society and industry. Computer **50**(9), 18–28 (2017)
8. Nakamoto, S.: Bitcoin: a peer-to-peer electronic cash system (2008). http://bitcoin.org/bitcoin.pdf
9. Phillip, A., Chan, J., Peiris, S.: A new look at cryptocurrencies. Econ. Lett. **163**, 6–9 (2018)
10. Buterin, V., et al.: A next-generation smart contract and decentralized application platform. White paper (2014)
11. Gray, M.: Introducing project "bletchley" - current blockchain ecosystem and evolution. https://github.com/Azure/azure-blockchain-projects/blob/master/bletchley/bletchley-whitepaper.md#current-blockchain-ecosystem-and-evolution, (2017). Accessed 05 Oct 2018
12. Gambetta, D., et al.: Can we trust trust. Trust.: Mak. Break. Coop. Relat. **13**, 213–237 (2000)
13. Resnick, P., Zeckhauser, R., Swanson, J., Lockwood, K.: The value of reputation on eBay: a controlled experiment. Exp. Econ. **9**(2), 79–101 (2006)
14. Resnick, P., Kuwabara, K., Zeckhauser, R., Friedman, E.: Reputation systems. Commun. ACM **43**, 45–48 (2000)
15. von Reischach, F., Michahelles, F., Schmidt, A.: The design space of ubiquitous product recommendation systems. In: Proceedings of the 8th International Conference on Mobile and Ubiquitous Multimedia, MUM 2009, New York, NY, USA, pp. 2:1–2:10. ACM (2009)

Privacy Centric Collaborative Machine Learning Model Training via Blockchain

Aman Ladia$^{(\boxtimes)}$

Liquid Protocol, Mumbai, India
aman@amanladia.com

Abstract. This paper tackles the issue of data siloing, where organisations are unable to share data with each other because of privacy concerns. Machine Learning models, which could benefit greatly from larger data sets shared between organisations, suffer in this era of data isolation. To solve this problem, a blockchain based implementation is proposed that allows training of machine learning models in a privacy compliant way. Instead of using blockchain in a typical database-style manner, the proposed solution uses blockchain as a means to handle joint ownership and joint control over a computer system known as the Training Machine. The Training Machine, set-up jointly by consortium members, serves as a secure, independent container that accepts data sets and an untrained model as inputs from different entities, trains the model internally, and outputs the trained model without revealing any data to other entities. Data is then deleted automatically. Blockchain ensures that this machine is not under the control of any one entity but is rather controlled transparently by all data-sharing parties. By placing sensitive information in an isolated system, and establishing blockchain based access control, the solution ensures that data is not accessible to any party other than the owner. The paper also shares use cases of this technology, along with a risk analysis and proof of concept.

Keywords: Private data sharing · Shared model training ·
Blockchain access control · Consortium data exchange · Deep learning training

1 Introduction

In recent years, Machine Learning (ML) has gained popularity as a data analysis technique. Central to most ML applications is the construction of an algorithm that allows a program to read existing data, formulate a mathematical expression (or 'model') and make predictions without human interference [1]. The accuracy of such a system relies heavily on the quantity of data that is supplied to it [2]. Collectively, there is more than enough data to build highly accurate models, but this data is scattered in 'silos': individual, isolated data stores that are disconnected from each other and accessible only to the company that operates them [3]. Privacy regulations like GDPR have only widened this disconnect [4]. As explored in Sect. 5, companies could benefit tremendously if they could somehow train their ML models using the combined knowledge of their data silos [5]. Unfortunately, in the absence of a transparent,

© Springer Nature Switzerland AG 2020
J. Prieto et al. (Eds.): BLOCKCHAIN 2019, AISC 1010, pp. 62–70, 2020.
https://doi.org/10.1007/978-3-030-23813-1_8

accountable, privacy-centred training protocol for ML systems, data will continue to be fragmented and the benefits of shared ML training will not be reaped.

Consider that a set of organisations (companies) operating in the same industry (e.g. banking or shipping) are members of a consortium (an association of companies). One of the companies has constructed an ML model – for instance, one that predicts the risk of credit default based on spending history and some demographic factors. Looking at the data of its own customers, the bank finds a strong concentration of data points in a relatively narrow range of variables: say middle aged, Caucasian men earning less than $50,000 a year. With little data for other demographics, the trained ML model yields inaccurate results for the minority cohort.

This model, however, can be improved by training it on data sets provided by other banks in the consortium who have clientele in the minority demographic.

To enable this cross-training of ML models on external data sets, this paper presents a blockchain based privacy centric method for training ML models on shared data points without the private data being exposed to a third party. The contributions of this paper are:

- It describes the construction of a blockchain network responsible for handling access-control to a dedicated, jointly-owned computer system responsible for training data in a digitally inaccessible environment (Sect. 3)
- It details the flow of transactions over the blockchain network that ensure privacy for all parties involved in the data training operation (Sect. 4.2)
- It delineates a proof-of-concept implementation of the same (Sect. 4.3), an associated risk analysis (Sect. 4.4) and industrial use cases (Sect. 5).

2 Problem Statement

Organizations should be able to share data for ML model training without the data being revealed to any party other than the owner.

However, they are unable to do so because existing solutions require a trusted party. It is difficult to monitor how data is used once it has been transferred to another party, making the process opaque and exposing companies to a high risk of data misuse/ theft. Consequently, companies face two extremes with no middle-ground:

- If trusted parties are used for data sharing, there are very few precautions organisations can take to ensure data is used only for the purpose of training ML models. With strict regulations like GDPR in place, a single incident of data misuse can lead to heavy fines for companies, withdrawal of permits and public disgrace (high risk).
- If data sharing is stopped entirely, the ML models produced will be of inferior quality and may perform poorly when compared to those models trained on larger data sets. This could lead to losses in cases where models make business-critical decisions.

This paper provides the middle-ground by proposing a blockchain-based system that ensures that (a) data is never transferred to a third company, and is not visible to any party other than the owner and (b) provides a transparent audit mechanism to ensure that the processing of the data is in complete control of the members of the consortium.

3 Overview

Accomplishing the two criteria mentioned in the problem statement requires the setup of a container that can train the ML model of one company (henceforth referred to as 'entity') on the data supplied by others, without any party seeing the other's data. This container runs on a designated 'Training Machine' (TM). The TM is a set of computers owned jointly by all members of the consortium. Each and every member holds equal access rights to the TM, but no member can access it independently without consensus from the other members. This paper builds such an access control mechanism using blockchain, where any access requests must be processed by all consortium members and require consensus to be accepted. Even then, direct shell access is not provided – only an interface to supply training data and the untrained model for processing is made available.

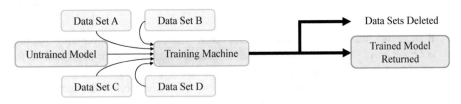

Fig. 1. Training Machine as a closed container

Effectively, this makes the TM a closed container which accepts data sets and an untrained model as inputs, trains the model internally, and outputs the trained model to its owner without exposing any data in the process (depicted in Fig. 1). This makes the entire training process autonomous and removes the need for a trusted third party. For added transparency, any changes made to the TM are automatically broadcasted to a blockchain network to ensure that any unauthorized action is open for audit.

To clarify, no data is transacted on the blockchain itself - doing so would open up the risk of malicious nodes storing copies of the information. Instead, data is directly transferred to the TM using traditional methods, while the blockchain handles any and all access requests with the additional security of transparent audit logs.

Additionally, this paper utilizes private channels to allow specific members of the consortium to share data. A Hyperledger Fabric channel is a private "subnet" of communication between two or more specific network members, for the purpose of conducting private and confidential transactions [6]. This means that even if only a few organisations wish to share data (e.g. 5 out of 15 consortium members), they can do so. This makes this solution highly modular and reusable – and independent chain does not need to be set up every single time.

4 System Design and Architecture

4.1 System Components

The first component of the proposed solution is a private, permissioned Hyperledger Fabric style blockchain called the Data Control Chain (DCC). All members of the consortium exist as organisations on the DCC, running independent peers and ordering services. The DCC is the primary blockchain channel, and the only one that can authorize access requests. Each peer on the DCC has the necessary chaincode installed required to vote in favour or against an access request, submit a training proposal, send an untrained model to the training machine and provide the access path to test data. These contracts are explained further in Sect. 4.2.

The second component is the Training Machine (TM), which must run a software called the 'Ledger Bridge', built to the following specifications:

- The Ledger Bridge should be able to read commands issued on the DCC/private channels and execute training operations accordingly.
- It should manage user access by creating temporary access accounts and editing the SSH key list file.
- It should monitor processes and dump errors/warnings on the DCC.

The last component is private communication channels between two or more entities willing to collaborate data for model training. These channels are instantiated within the Hyperledger Fabric network when some, but not all parties wish to participate in a data sharing operation.

4.2 Transaction Flow

Let us assume there is a consortium that wishes to make use of private data sharing for ML training, via blockchain. First, the consortium must purchase and set up the TM. The TM is an array of computers of sufficient computational capacity need to carry out ML training operations. The physical location of the TM can be jointly decided. Digitally, the TM runs a Docker container with a Hyperledger Fabric node that is connected to a pre-established DCC among consortium members. It also runs the Ledger Bridge (Sect. 4.1) and any software needed for training of ML models (e.g. Python) with associated dependencies (e.g. Tensorflow). The initial setup can be performed under the supervision of representatives from all consortium members.

The set up only needs to be completed once. Once initialised, the TM should be running autonomously, under the sole control of the Ledger bridge which in turn is controlled through the DCC.

For simplicity, let us consider a scenario where out of a consortium of n companies, two entities, A and B, wish to share training data for a model developed by entity C. They first sign an off-chain legal agreement and then proceed with the steps detailed next (Fig. 2).

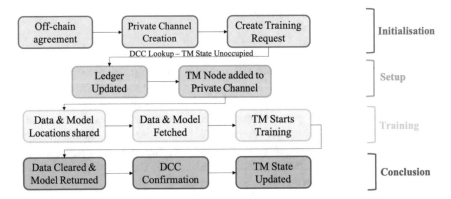

Fig. 2. Process for data sharing between *n* parties. This phase can be divided into four stages: (1) Initialisation, (2) Setup, (3) Training and (4) Conclusion

1. Entities A, B and C create a private channel
2. The ML model owner (entity C in this case) executes the `CreateTrainingRequest` smart contract on the DCC. The parameter specified is the name of the private channel created earlier. This contract performs a lookup of the DCC to check the state of the TM: whether it is occupied with an ongoing operation or not. If not, the request passes, and the ledger is updated to reflect the channel which the TM needs to join.
3. The TM joins the private channel. Entities A and B run the `AddData` smart contract on the private channel, which specifies the locations of the data to be fetched.
4. Finally, the `RunRequest` contract is called by the model owner, specifying the location of the model files, and the return location of the trained model. This signals the Ledger Bridge to start the training process.
5. The Bridge fetches the data and the untrained model from the locations specified. Note: both these locations are on the independent servers of entity A, B and C, and it is assumed that all entities have access control systems (possibly Public Key Infrastructure) in place that only allow GET requests from the Training Machine.
6. The Bridge runs the model training files on the data sets.
7. Once training has elapsed, the data sets are wiped by the Ledger Bridge, and upon successful deletion, a confirmation is posted on the DCC for transparent audit.
8. The trained ML model is returned to entity C. It is also deleted, and a confirmation is posted for open audit as before.
9. DCC is updated to indicate the TM is free for a new operation.

There is also a provision to allow entity C to gain direct access to check, update and maintain the training process. In order for such access to be granted, a proposal must be raised on the DCC, listing out the public key of the accessor, the period of access, and privileges needed. All members vote on the request; if it passes, the DCC state is updated. The Ledger Bridge creates a limited access user profile and adds entity C's cryptographic fingerprint to the SSH key list of the Training Machine for temporary access. Any changes made to the file system are recorded dumped onto the DCC for transparent audit. After access time elapses, access is revoked automatically (Fig. 3).

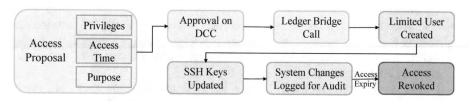

Fig. 3. Flow diagram for requesting access rights to the Training Machine

4.3   Proof of Concept

This protocol was implemented to demonstrate a proof-of-concept and not present a final implementation; hence, the setup was simplified such that it could be quickly deployed and an assessment could be performed without complications. It consisted of a Hyperledger Fabric network with three organisations, each with one peer and one orderer. Each organisation was simulated on a separate computer. A fourth computer was added that served as the Training Machine (specifications here: [7]). A TensorFlow program that consisted of a basic Linear Regression algorithm was chosen as the test model. An open source data set of Average Income and Crime Rate for 2000 cities in the United States was obtained from UCI ML Repository and used for training [8].

This experimental setup recreated a consortium with three entities willing to share data. It was assumed that all three entities wish to pool in data for a particular model, and hence there was no need for a private channel - all operations occurred over the DCC itself. All four systems (3 consortium members and 1 TM) were added to the DCC. The TM was set up with a Python bridge that could access the DCC ledger. As prescribed, the TM was allowed to run autonomously after initial setup was performed.

In the test setup, the execution of the training finished in a very short time (a few seconds) because of the small data set. Once complete, the model was appropriately returned and data traces deleted. The ledger state was updated to reflect finished training and the deletion of data files. The time taken to implement this small-scale network was relatively minimal. Industrial robust development would take longer, but this is a one-time establishment cost. Hence, the barrier to entry exists largely at the implementation level, not the operational level. Overheads associated are more to do with the model training component and less with the blockchain access control mechanism. Fortunately, this system is modular, and the ML training component can be modified, for instance, to a distributed setup, according to the technical needs of the industry.

4.4   Risk Assessment

A qualitative risk assessment of this technique was performed. The following are the potential risks associated with the implementation of the system described in this paper, along with a brief analysis of the likelihood of risk:

- Majority of Nodes Compromised: If more than 50% of DCC nodes are compromised simultaneously, data integrity is at risk. However, this is highly impractical in a large consortium with industry-grade security. Improbable in an enterprise environment.
- Data fetch operation compromised: Data is fetched directly by the TM from the data sharing entities' servers. Access requests at the server end are authenticated using Public Key Infrastructure. Improbable chances of interception with secure (encrypted) copy operations.
- Ledger bridge compromised: Ledger bridge implementation must be strong. Enterprise programs are scrutinized at every level, so the likelihood is remote.
- Malicious code in ML model/training data: Ill-intentioned organizations can embed malicious code in the model or data. Sandboxing the operation and running suitable malware checks can resolve the problem. Remote risk.

On a scale of 1 to 5 (as indicated in the risk assessment matrix in [9]), the risk can be rated at a 1.5, as it lies between 'improbable' and 'remote'. The severity of the impact is significant. A data breach can lead to backlash and legal complications. However, with the implementation of audit measures that keep the consortium informed against any unauthorized operations, the risk can be mitigated. Hence, the severity of the Risk can be rated between 3 (moderate) and 4 (significant), approximately 3.5. The combined risk rating is 5.25, putting it in the green, acceptable zone of the risk matrix.

5   Application Use Cases

A potential intra-industry use case of such a system would be banks sharing credit card fraud data to improve risk analysis algorithms. In this example, banks belonging to a consortium (e.g. R3 blockchain development consortium) could set up a data sharing system and then use the medium to build and improve on ML risk analysis models [10].

Similarly, the insurance sector can pool data to produce better liability estimates and set more accurate premiums. The insurance industry as a whole—not just independent companies—can benefit from such an arrangement as even a marginally more accurate analysis could save the industry millions of dollars. Customers are also likely to benefit: part of the savings can be passed on as lower premiums.

By extension, the private sharing system could be extended to allow the creation of a 'super-classifier' that is built by combining individual classifiers from different banks. Model ensembling (combining two or more ML classifiers to arrive at a more accurate classifier) techniques like stacking can be incorporated into the system. Here, instead of taking untrained algorithms and test data as inputs, the system could accept trained classifiers along with random test data. It could then apply ensemble learning techniques to arrive at a final model, which can be shared amongst participating entities.

6 Limitations

The proposed system requires an amalgamation of blockchain with traditional centralised processing. Although efforts have been made to implement this in the most secure way, some limitations exist.

Maintenance of the Training Machine is presently a manual operation. A consortium member can only request limited access rights over the DCC. Such an access does not have administrative rights, which are required by some operations (e.g. updating training software: `apt-get update` requires `sudo`). Currently, an admin operation requires the manual supervision of representatives from consortium members (as decided in a legal off-chain agreement). In the future, a provision can be made for admin commands to be transparently executed over the DCC with member consensus.

Linked to first limitation is one to do with choice of platform. Currently, the implementation supports one ML platform and library (Python TensorFlow in the Proof of Concept). As different applications require different platforms, the capabilities of the Ledger Bridge must be expanded to include provisions for multiple platforms.

7 Conclusion

In this paper, a framework for the private training of Machine Learning models was described.

The requirements of such a system were summed up to an assurance of data privacy, model privacy, and a transparent audit mechanism. The implementation of the system involved the construction of a Hyperledger Fabric Data Control Chain (DCC), along with the acquisition of a jointly owned Training Machine and the provision for private channels between members. The DCC was designated as the main chain on which any access requests to the Training Machine would be handled. The implementation of a Ledger Bridge was also delineated, such that model training on the Training Machine could be controlled by chain code.

A proof of concept was provided, and potential use cases of this system in an industry setup were identified. Significant financial benefit was also illustrated via use cases. Technical limitations and risks associated with the setup were analysed, and some extensions were suggested to make the system even more robust in the future.

References

1. Domingos, P.M.: A few useful things to know about machine learning. Commun. ACM **55** (10), 78 (2012). JCotA
2. Witten, I.H., Frank, E., Hall, M.A., Pal, C.J.: Data Mining: Practical Machine Learning Tools and Techniques. Morgan Kaufmann (2016)
3. Tene, O., Polonetsky, J.: Big data for all: privacy and user control in the age of analytics. Nw. J. Tech. Intell. Prop. **11**, xxvii (2012)
4. General data protection regulation (2016). 2016/679

5. Mougayar, W.: The Business Blockchain: Promise, Practice, and Application of the Next Internet Technology. Wiley, Hoboken (2016)
6. Cachin, C.: Architecture of the hyperledger blockchain fabric. In: Workshop on Distributed Cryptocurrencies and Consensus Ledgers (2016)
7. Implementation specifications. http://liquidprotocol.io/specs.pdf. Accessed 24 Mar 2019
8. Communities and crime data set. https://archive.ics.uci.edu/ml/datasets/Communities+and+Crime. Accessed 24 Mar 2019
9. Risk matrix. http://liquidprotocol.io/risk_matrix.png. Accessed 24 Mar 2019
10. Galindo, J., Tamayo, P.: Credit risk assessment using statistical and machine learning: basic methodology and risk modeling applications. Comput. Econ. **15**(1–2), 107–143 (2000). JCE

Fuzzy Rules Based Solution for System Administration Security Management via a Blockchain

Arnaud Castelltort[1], Antoine Chabert[2], Nicolas Hersog[2], Anne Laurent[1(✉)], and Michel Sala[1]

[1] Univ Montpellier, LIRMM, CNRS, Montpellier, France
{arnaud.castelltort,anne.laurent,michel.sala}@umontpellier.fr
[2] ChainHero, Montpellier, France
antoine.chabert@tohero.fr, nicolas.hersog@chainhero.io
http://www.lirmm.fr
http://www.chainhero.io

Abstract. Digital transformation has led to the fact that almost all organizations and companies are provided with internal private networks and manage sensitive data and applications. In this context, system administrators are superusers who can access all this sensitive material. As it is known that many frauds are caused by internal actions, we argue that it is important to be provided with strong automated logging systems even for superusers. For this purpose, blockchains are an efficient solution as they cannot be overwritten by the system administrators. However, as it is not efficient to store all the actions, we introduce in this paper a novel system based on fuzzy rules in order to efficiently manage the system logging system in a blockchain.

Keywords: Blockchain · Fuzzy rules · System administration · Log files · Fraud detection

1 Introduction

As the organizations are more and more using digital solutions to manage their internal processes and data, sensitive material is more and more managed by the information system. Security is thus a critical issue and is a major concern and expense. Information systems and infrastructures are thus protected and all operations are logged in journals. However, it has been reported by the U.S. State of Cybercrime that *"23% of electronic crime events were suspected or known to be caused by insiders"* and that *"45% of the respondents thought that damage caused by insider attacks was more severe than damage from outsider attacks [...] customer records compromised or stolen, confidential records (trade secrets or intellectual property) compromised or stolen, and private or sensitive information [...] unintentionally exposed"* [10]. As system administrators have access to sensitive information without any restriction as being *root* on the systems,

© Springer Nature Switzerland AG 2020
J. Prieto et al. (Eds.): BLOCKCHAIN 2019, AISC 1010, pp. 71–78, 2020.
https://doi.org/10.1007/978-3-030-23813-1_9

it is important to be able to monitor their operations so that they do not take benefit from their position while ensuring to cover up their activities.

For this reason, the ChainHero company has built the Blockaudit system that allows to report in a blockchain the system administrator operations. This prevents from any falsification. The architecture is described in Fig. 1, the smart engine being responsible for choosing the operations being logged in the blockchain.

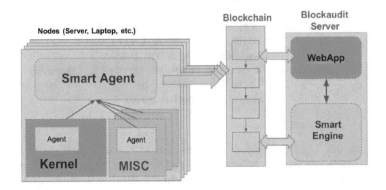

Fig. 1. Current crisp architecture

The Hyperledger Fabric private blockchain has been chosen by the company. The blockaudit agents are spread over the network on the critical servers and report all actions on the blockchain, using the *auditd* tool developed by RedHat and delivering information from the Linux kernel. Every agent is provided with predefined behaviours so as to determine which operations are reported, e.g., file access or network operations. Rules can be added so as to follow up specific operations, as for instance determining softwares, users or files to be taken into account. Such operations are then collected and stored in the blockchain so as to be reported for users. Rules can be associated with the level of criticality.

However, these rules are too crisp to be easily applied as the operation needs to perfectly feet the rule so as to be detected as critical, which may lead to problems.

For instance, if getting connected 5 times on a server in the night is considered as critical then an ill-intentioned person who would connect only 4 times or one minute after the night would not be discovered.

In this work, we have thus proposed to soften the rules by relying on fuzzy rules. Our solution has been implemented and tested.

2   Related Works

2.1   System Administration and Audit

Auditing systems is a key element for managing the security [12]. Several risks are associated with such insiders [5]: data leaks, access control,... Data leakage can be of different forms and data leakage prevention systems (DLPSs) have been designed [1]. Some of them are focused on internal threads, as in [8,9].

In this work, we focus on the operations close to the Linux kernel. For this purpose, we focus on auditd which is the userspace component to the Linux Auditing System and is responsible for writing audit records [11].

In our work, we rely on fuzzy rules, described below.

2.2   Fuzzy Rules

When dealing with rules, the systems are based on a set of *if ... then...* statements where the antecedent is either true or false. However, in some cases, it is important to soften this in order to consider antecedents that are partially true. For this purpose, the fuzzy logic framework [3] is relevant. In such a framework, objects gradually belong to the fuzzy subsets with a membership degree ranging from 0 to 1. Every fuzzy set is defined over a universe and is associated with a fuzzy membership function that allows to compute this degree.

Fuzzy rules are then of the form *if A then B* where A and B are fuzzy sets, as for instance *if the number of connections is high then the level of criticity is low*. Multiple variables can be aggregated by using t-norms and t-conorms, as for instance *if the number of connections is high AND the period is night then the level of criticity is high* where *low, high* and *night* are fuzzy concepts described by their membership function.

Figure 2 describes a fuzzy membership function for the concept of night, the universe being the hours (from 0 to 24) in a day.

Fig. 2. Fuzzy membership function describing night and day

Fuzzy quantifiers can be defined in the same manner in order to describe quantities and percentages with words, as for instance *many, almost 5, few, etc.*

Fuzzy concepts are used in our approach in order to describe the concepts appearing in the rules in a soft and understandable manner as crisp partitions are often impossible for users to define.

It has already been shown that it is interesting to link blockchain and artificial intelligence [7], and fuzzy logic [4,6], as it is the case for cyber security [2]. But no work has coupled blockchains and fuzzy rules as proposed in our work and described below.

3   FBA: The Fuzzy BlockAudit Approach

Our work aims at defining a solution to store the operations of the system administrators of critical servers in a secure manner. We thus rely on blockchains. The operations are taken from the *auditd* tool on Linux kernel. As we claim that it is not necessary to store all operations, we define rules that determine what operations deserve to be written in the blockchain. These rules depend on so-called *contexts* as described below.

3.1   Contexts

Contexts are used in the definition of fuzzy variables. For example, *connection* and *command* have been implemented in our solution: they are used to define the fuzzy sets of *low* and *high* for both contexts (connection and command). The fuzzy variables are then used to:

– determine if a rule must be triggered;
– retrieve the information from the event;
– compute to which extent the premise of a rule is matched;
– update the premises of the rules.

For instance, in the *connection* context, the value 10 represents a *high* number of connections whereas this value is rather *low* in the *command* context.

3.2   Fuzzy Rules

These fuzzy variables are then used in fuzzy rules. Rules are of the form:

```
RULE RuleName [ Expression ] AS CriticalityType
```

where expression is recursive and is composed of other expressions connected with logical AND, OR, XOR operators. In a rule, the premise is of the form

```
AMOUNT (context) SPEED FROM (ACCESS -> ACCESS) AT (PERIOD)
```

where:

– AMOUNT (context) is the number for triggering the rule for the given context.
– SPEED allows to follow up the events over time.
– AT allows to restrict the period.

For instance, a premise may be only valid for holiday periods. It should be noted that the definition of the periods is not easy as nights overlap 2 days and nights, and cannot be defined as a single subset of the hours between 0 and 24 using modulo 24.

A critical level is also considered. *WEAK* critical level (as for instance *cd* commands, read and write on non sensitive files) are distinguished from

NORMAL critical level (as for instance change of user, access to a shared folder), *STRONG* critical level (as for instance the mounting of a folder, the use of startup script), and *CRITICAL* (as for instance reboot, change of audit rules, adding or deleting a user account, read and write on sensitive files).

Depending on the level of criticity, the events on premises are aggregated using various operators, as listed in Figs. 3 and 4.

Criticality	T-Norm
WEAK	f(a,b) = a * b
NORMAL	f(a,b) = 0.7 * min(a,b) + (1 - 0.7)(a + b)/2
STRONG	f(a,b) = 0.8 * min(a,b) + (1 - 0.8)(a + b)/2
CRITICAL	f(a,b) = min(a,b)

Fig. 3. Aggregating degrees with AND operator (t-norm)

Criticality	T-Conorm
WEAK	f(a,b) = max(a,b)
NORMAL	f(a,b) = 0.8 * max(a,b) + (1 - 0.8)(a + b)/2
STRONG	f(a,b) = 0.7 * max(a,b) + (1 - 0.7)(a + b)/2
CRITICAL	f(a,b) = 1 - (1 - a)(1 - b)

Fig. 4. Aggregating degrees with OR operator (t-conorm)

4   Architecture and Implementation

Our solution is based on the management of the operations that trigger fuzzy rules as described previously if they are considered as worth being stored in the blockchain. The fuzzy rules are defined thanks to a DSL.

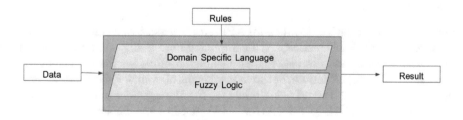

Fig. 5. Current architecture

4.1 Domain Specific Language

As the rules must be easily configured by using a simple language, we use an external DSL that allows to define the rule with the following grammar.

```
RULE identifier
OL (logical operator): [\AND", \OR", \XOR"]
Parenthesis: [\(\, \)"]
Expression <- (Expression) OL Premise
Expression <- Prémisse OL Expression
Premise:
AMOUNT: [\HAPPEN", \FEW",\MODERATE", \ MANY"]
SPEED: [\SLOWLY", AVERAGE", \QUICKLY"]
ACCESS: [\USER", \ROOT", \SYSTEM"]
PERIOD : \AT(\ [\NIGHT", \OFFICE", HOLIDAYS", \WEEKEND"] \)".
CRITICITY
 [\AS"] [\WEAK", \NORMAL", \STRONG", \CRITICAL"]
```

A rule always starts with

```
[\RULE"]
```

An example is given by Fig. 6.

```
RULE exempleRule [
    FEW(connection) QUICKLY FROM(USER -> ROOT) AT(NIGHT)
    OR
    MANY(command) SLOWLY AT(NIGHT)
]
AS CRITICAL
```

Fig. 6. Example of a rule

4.2 Architecture

Our solution is accessible through an API (Application Programming Interface) exposed on the HTTP protocol. All components are easily configurable by the users with several modules (see Fig. 8): ContextsModule for the management of contexts, DSL, ParserModule for the management of events, RulesModule for the management of rules, FuzzyLogicModule for the management of fuzzy operators. A document-based NoSQL storage backend is used as an efficient storage for raw textual data from JSON that are pre-treated.

One of the key aspect of this software architecture is the ability to be extended with context modules. Rogue behaviors are behavioral anomalies that can occur in human activities and that can thus be retrieved from human generated data. The context modules allow the user expert to override some predefined rules to adapt them to a specific context and also to extend the system with new or domain specific fuzzy terms, rules and then ultimately more rogue behavior definitions.

Figures 7 and 8 show the architecture and technologies being used.

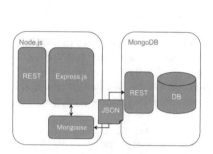

Fig. 7. Technical architecture of the solution

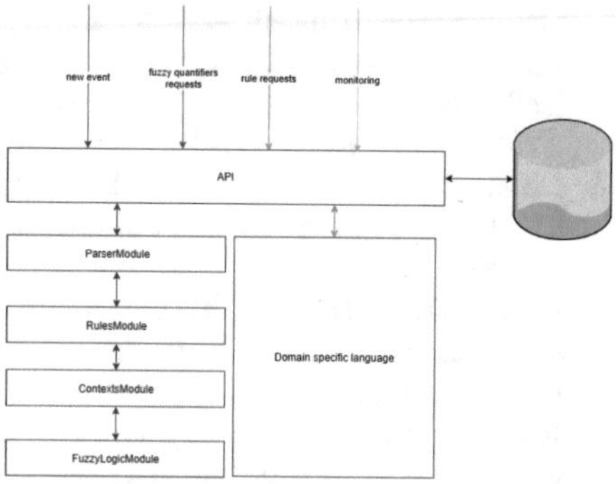

Fig. 8. Architecture of the solution

The solution is easily pluggable in an environment with an HTTP event-based application server such as NodeJS server. It is possible to dynamically configure, add, and delete rules through the API. It does not require to reboot and restart which could allow the system to adapt itself while running (which is a further work). Data visualization and monitoring systems are available allowing the user to take decisions.

5   Conclusion

In this paper, we present an original method for managing system administrator operations logging in a blockchain by the use of fuzzy rules. This approach allows companies to prevent system administrators to take benefit from their position for attacking critical servers while ensuring to cover up their activities and to erase their traces. The solution has been implemented, tested and is in deployment in the information system of a confidential client.

Our future work includes the automatic learning of rule evolutions and the monitoring of operations outside the Linux kernel monitored by auditd.

Acknowledgments. The authors would like to thank Loïc Combis, Kevin Hassan, and Hugo Maitre, students from Polytech Montpellier, for their help for implementing and testing the approach.

References

1. Alneyadi, S., Sithirasenan, E., Muthukkumarasamy, V.: A survey on data leakage prevention systems. J. Netw. Comput. Appl. **62**(C), 137–152 (2016)
2. Al Amro, S., Chiclana, F., Elizondo, D.A.: Application of fuzzy logic in computer security and forensics. In: Elizondo, D.A., Solanas, A., Martínez-Ballesté, A. (eds.) Computational Intelligence for Privacy and Security, vol. 394. Studies in Computational Intelligence, pp. 35–49. Springer (2012)
3. Bellman, R.E., Zadeh, L.A.: Decision-making in a fuzzy environment. Manag. Sci. **17**(4), B-141–B-164 (1970)
4. Chen, R.-Y.: A traceability chain algorithm for artificial neural networks using T-S fuzzy cognitive maps in blockchain. Futur. Gener. Comput. Syst. **80**, 198–210 (2018)
5. Homoliak, I., Toffalini, F., Guarnizo, J., Elovici, Y., Ochoa, M.: Insight into insiders: a survey of insider threat taxonomies, analysis, modeling, and countermeasures. CoRR (2018). arXiv:abs/1805.01612
6. Kaga, Y., Fujio, M., Naganuma, K., Takahashi, K., Murakami, T., Ohki, T., Nishigaki, M.: A secure and practical signature scheme for blockchain based on biometrics. In: Liu, J.K., Samarati, P. (eds.) Information Security Practice and Experience - 13th International Conference, ISPEC 2017, Melbourne, VIC, Australia, 13–15 December, 2017, Proceedings. LNCS, vol. 10701, pp. 877–891. Springer (2017)
7. Marwala, T., Xing, B.: Blockchain and artificial intelligence. CoRR (2018). arXiv:abs/1802.04451
8. Spitzner, L.: Honeypots: catching the insider threat. In: Proceedings of the 19th Annual Computer Security Applications Conference, ACSAC 2003, p. 170. IEEE Computer Society, Washington, DC (2003)
9. Stolfo, S.J., Salem, M.B., Keromytis, A.D.: Fog computing: mitigating insider data theft attacks in the cloud. In: Proceedings of the 2012 IEEE Symposium on Security and Privacy Workshops, SPW 2012, pp. 125–128. IEEE Computer Society, Washington, DC (2012)
10. Trzeciak, R.F.: SEI cyber minute: insider threats (2017). http://resources.sei.cmu.edu/library/asset-view.cfm?assetid=496626
11. Xu, H., Tang, R.: Study and improvements for the real-time performance of Linux kernel. In: 2010 3rd International Conference on Biomedical Engineering and Informatics, vol. 7, pp. 2766–2769, October 2010
12. Zhao, K., Li, Q., Kang, J., Jiang, D., Hu, L.: Design and implementation of secure auditing system in Linux kernel. In: 2007 International Workshop on Anti-Counterfeiting, Security and Identification (ASID), pp. 232–236, April 2007

A Balanced Routing Algorithm
for Blockchain Offline Channels
Using Flocking

Subhasis Thakur[✉] and John G. Breslin

National University of Ireland, Galway, Ireland
{subhasis.thakur,john.breslin}@nuigalway.ie

Abstract. Offline channels have the potential to mitigate the scalability problem of blockchains. A Path-Based fund Transfer (PBT) uses a path in the channel network. PBTs can make the channel network imbalanced, i.e., funds in a few channels become very low and funds in other channels become very high. Imbalanced channel network may make PBT infeasible. Hence we need a routing algorithm for PBT that keeps the channel network balanced. In existing solutions for this problem have privacy problem as the channels have to reveal their balances in order to find suitable routes for PBTs. In this paper, we mitigate this problem as we propose a flocking based algorithm for PBTs that keeps the channel network balanced.

Keywords: Offline channels · Bitcoin lightning network · Flocking

1 Introduction

Public and proof of work based blockchains have a scalability problem. For example, the Bitcoin network can process 7 transactions per second, Ethereum network can process 15 transactions per second while Mastercard can process 50000 transactions per section. Offline channels can improve the scalability of blockchains. Offline channels such as Bitcoin lightning network aims to reduce the number of transaction to be recorded in the blockchain by allowing offline transactions among pairwise parties. An offline channel among two parties aims to create only two transactions in the blockchain. One transaction is created during opening the channel and another transaction is created to close the channel. The blockchain does not require to remain informed about transactions between these two events. A Path-Based fund Transfer (PBT) in channel network uses a path in the channel network to transfer fund between two peers who do not have a mutual channel.

There are few problems with the blockchain offline channels that prevent its practical deployment for micropayments. For greater longevity of the channel network (maximum usage of channel network without recording transactions into the blockchain), PBTs must be coordinated to maintain the balance of a channel

© Springer Nature Switzerland AG 2020
J. Prieto et al. (Eds.): BLOCKCHAIN 2019, AISC 1010, pp. 79–86, 2020.
https://doi.org/10.1007/978-3-030-23813-1_10

network. Balancing the channel network means maintaining uniform usage of channels. For example: Let the pair of peers A and B (and the pair of peers B and C) have bidirectional channels between them. If the path $A \rightarrow B \rightarrow C$ is used for fund transfer between A and C then, channels $A \rightarrow B$ and $B \rightarrow C$ will lose fund and mirrored channels ($B \rightarrow A$ and $C \rightarrow B$) will gain funds. The repeated usage of the paths $A \rightarrow B$ and $B \rightarrow C$ will drastically reduce their value and eventually they will no longer support PBT. The longevity of a channel network can be improved by coordinating PBTs. The current methods for such coordination collect PBT information to predict the best paths for future PBTs that maintain a balanced channel network. The problem with such an approach is the privacy of the peers. Peers must reveal their fund transfer information to maintain the balance of the channel network. In this paper, we propose a routing algorithm that preserves the privacy of peers as they do not need to reveal PBT information to maintain balanced channel network. The paper is organized as follows: In Sect. 2 we mention related literature and in Sect. 3 we describe the basics of offline channels. In Sect. 4 we present our routing algorithm. We evaluate our solution in Sect. 5 and we conclude the paper in Sect. 6.

2   Related Literature

Bitcoin lightning network was proposed in [9] which allows peers to create and transfer funds among themselves without frequently updating the blockchain. Similar networks are proposed for Ethereum [1] and credit networks [7]. A routing algorithm for Bitcoin lightning network was proposed in [10]. A fast routing protocol was proposed in [2]. A method for anonymous payment to improve privacy in PBT was developed in [4]. [5] proposed a decentralised routing algorithm for the channel network. [8] proposed privacy preserving routing protocol for PBT that can handle concurrency. [6] proposed a balanced routing algorithm for the channel network. [4] proposed privacy preserving schemes for fund transfer in the offline channel network. [3] proposed a privacy-preserving routing method for fund transfer in the channel network. The contributions in this paper advance the state of the art as follows: (a) Majority of routing algorithms for PBT in channels do not consider the problem of balancing the channel network. (b) Privacy-preserving solutions in these routing algorithms are not designed for maintaining privacy for balancing the channel network, rather these algorithms ensure the privacy of the sender and the receiver of a PBT. (c) [6] proposed a routing protocol that balances the network. It requires finding cycles in the channel network in order to make fund transfers to keep the channels balanced. Our proposed method does not need any fund transfer among the channels to keep it balanced. Also, this routing algorithm ignores the privacy of peers as they have to reveal the PBT information.

3   Offline Channels

In this paper, we use the offline channel network construction for Bitcoin lightning network [9]. The basic protocol for using an offline channel is as follows:

(1) Say Alice and Bob want to create a channel between them with balances 10 tokens (each contributes 5 tokens). (2) Alice and Bob create two pairs of lock (hash) and key (random string). They exchange the locks. (3) Bob creates a 'confirmation transaction' as follows: (3.a) There is a multi-signature address between them which requires a signature from both to transfer fund from it. We will call this address M_1. (3.b) Bob creates transactions from M_1 which states that Bob will get 5 tokens and remaining 5 tokens will go to another multisignature address between them. We will call this address M_2. (3.c) The 5 tokens in M_2 will be given to Alice after 10 days or Bob can claim it if it can produce the key to the lock of Alice. (4) Bob signs this transaction and sends it to Alice who can use it to get tokens from the channel by signing it and publishing it to the blockchain network. (5) Alice produces mirrored confirmation transaction and sends it to Bob. The confirmation transactions ensure that both parties can recover from if they fund the channel between them. (6) Now Alice and Bob transfer fund in the multi-signature address by creating transactions in the blockchain and hence the channel becomes operational. (7) Both parties should exchange keys and create new confirmation transaction to update the channel. (8) If any party announces a confirmation transaction then the channel closes.

The protocol for path based transfer in channel network is as follows: Say Alice wants to send fund to Carol via Bob. (1) Carol will create a lock and a key. (2) In the multi-signature address between Carol and Bob, a contract will be created as follows: (2.a) Bob will send 5 tokens to this address. (2.b) Bob will get these tokens back after 9 days if Carol does not claim it. (2.c) Carol can claim it anytime if it can produce the key to the lock. (3) Similarly, another contract will be created between Alice and Bob as follows: (3.a) Alice will send 5 tokens to this address. (3.b) Alice will get these tokens back after 10 days if Bob does not claim it. (3.c) Bob can claim it anytime if it can produce a key to the lock. (4) Thus Carol reveals the key to Bob as it collects the fund, which Bob uses to get refunded from Alice.

4   Flocking Based Routing Protocol

In this routing protocol, we preserve the privacy of channels by not gathering information on exact channel balances. Each channel is assigned a coordinate in two-dimensional space. In such space, channel distance between two peers is measured as the Euclidean length of the edge between them in the X-Y plane if they have a channel. We use the length of a channel to 'replace' value of a channel in order to preserve the privacy of the peers. A solution to keep the channels balanced is: increase the usage of channels with high value and decrease the usage of channels with low funds. In a similar approach, our routing algorithm prefers shortest paths (measured as the total Euclidean distance of a path) for PBTs. Our main innovation is how to keep updating distances of channels in order to reflect their usage. We solve this by using the flocking algorithm. Flocking algorithm mimicries how flocks of animals perform coordinated group traversal without crashing with each other only using local information about the position and velocity of animals in close proximity. We interpret flocking behavior to keep the channels balanced as follows:

(1) If a channel is used in a PBT then it moves in a specific direction according to the details of the PBT. The channels move in such a direction that mutual distances among the channels used in a PBT increase and the same for the complementary path decrease.
(2) Triggered by the movement of channels in a PBT, surrounding channels adjust their positions according to rules of flocking. Flocking behavior ensures equilibrium among channel distances. For example, as shown in Fig. 1 (right), a PBT uses path $V1$ to $V2$. Hence the distance between $V1$ and $V2$ is increased to make it less likely that it will be used again (less likely to be in any shortest path). But as $V1$ is moved to a new coordinate, the distance between $V1$ and $V3$ also increased. In order to compensate such increase in distance $V3$ will follow the flocking procedure and find a new coordinate to reduce their distance .

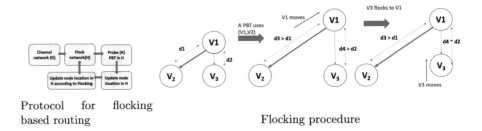

Protocol   for   flocking based routing

Flocking procedure

Fig. 1. Flocking based routing

The overview of the flocking based routing is shown in Fig. 1 (left). First, we generate the Flock graph from the channel network by creating a node every edge (a channel) in the channel graph. In the next section, we will assign coordinates to each node in the flock graph and describe the probing and PBT method for the Flock graph. After the execution of a set of PBTs, the flock graph is updated as the position of nodes are changed due to the execution of a PBT and due to flocking behavior. After updating the channel network with new coordinates for nodes in the Flock graph we execute the next set of PBTs.

4.1   Creation of Flock Graph H

Let $G = (V, E)$ be a directed graph representing the channel network with n peers (V) and m channels (E). $W(E) \mapsto \mathbb{R}^+$ be the value of the channels. We assume that the channels are bi-directional. Hence if $(V_i, V_j) \in E$ then there exists a complementary channel $(V_j, V_i) \in E$. We create a directed graph $H = (N, L)$ with nodes N and edges L from G as follows:

(1) For each edge $(V_i, V_j) \in E$ we create a node N_x in H. (2) Let $F \subset E$ be the set of edges incident on V_i. For each edge $(V_a, V_i) \in F$, we create an edge (N_y, N_x) in H where N_y is the vertex corresponding to the edge (V_a, V_i). (3) Let $F' \subset E$ be the set of outgoing edges from V_j. For each edge $(V_j, V_b) \in F$, we create an edge (N_z, N_x) in H where N_z is the vertex corresponding to the edge (V_j, V_b). (4) Vertex N_x's weight will be the same as the weight of the edge (V_i, V_j). (5) Each node N_x will be assigned a coordinate in the X-Y plane. Length of an edge in H will be determined by the Euclidean distance. $D(L_x)$ will denote the length of edge $L_x \in L$. (6) A path in H will denote the set of channels used in a PBT. The complementary path is the path in H corresponding to opposite channels in G.

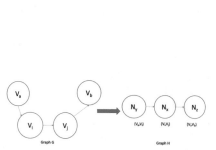

Creation of Flock graph H.

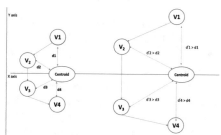

Nodes v_1, v_2, v_3, v_4 move away from the centroid after completion of the PBT $v1 \to v2 \to v3 \to v4$.

Fig. 2. Flocking procedure

4.2 Probing and PBT

Let V_x wants to transfer x tokens to V_y using a PBT. V_x uses the following steps:

1. Let $F \subset E$ be all outgoing edges from V_x and $F' \subset E$ be all incoming edges to V_y in the graph G.
2. Let $N^F \subset N$ be all nodes in the graph H corresponding to channels F and $N^{F'} \subset N$ be all nodes in the graph H corresponding to channels F'.
3. V_x can start the transfer from any node in the set N^F and end the transfer in any node in the set $N^{F'}$.
4. Thus V_x has to probe paths from the set N^F to the set $N^{F'}$.

Probing these paths are executed as follows:

1. Let $Cost$ be a $A \times B$ matrix where $|N^F| = A$ and $|N^{F'}| = B$. $Cost[i, j]$ will denote the distance starting from $N^F[i]$ to $N^{F'}[j]$.
2. $Cost[i, j]$ is calculated as the length of the shortest path in the graph H. Note that distances between the nodes in H are changing every time a new transaction is executed.
3. After computing $Cost$ matrix, V_x probs the paths starting from the lowest $Cost$ if it has enough balance to support its PBT.

4.3   Node Location Update in H

Coordinates of nodes in H are updated for every PBT. For each PBT $v1 \rightarrow v2 \rightarrow v3 \rightarrow v4$, we find the centroid of the v_1, v_2, v_3, and v_4. All nodes move further away from the centroid as shown in Fig. 2 (right). Similarly, nodes in the complementary path of the path $v1 \rightarrow v2 \rightarrow v3 \rightarrow v4$, move towards their centroid. In a general PBT algorithm, the sender of a PBT can inform all peers in the PBT to change their locations and in a multi-hop PBT, the sender can send such location change information to all peers by introducing small noise in the new location information.

4.4   Node Location Update in H Using Flocking

There are two steps in flocking [11] based node location change. Each node categories its surrounding nodes into two groups (1) Repulsion zone and (2) Align zone. Nodes in the repulsion zone are too close to a node and it wants to move away from them and nodes in align zone are neither too far from it nor too close to be in repulsion zone. Each node finds the mean angle of all nodes in its repulsion zone. The angle of a node is its heading calculated from the point $(0,0)$ in the given X-Y plane. The node finds the opposite of this mean angle by adding $180°$. Let's call this angle $Angle_1$. Similarly, the node finds the mean angle for all nodes in its align zone. Let's call this angle $Angle_2$. The position of the node (x_i, y_i) is changed as follows: $x_i = x_i + .2(Cos(Angle_1) + Cos(Angle_2)), y_i = y_i + .2(Sin(Angle_1) + Sin(Angle_2))$.

5   Evaluation

We use Bitcoin lightning network data. There are 2800 nodes and approximately 22000 edges. We create three subgraphs from this dataset excluding channels with a very high degree. We extract three subgraphs from the lightning network data with the number of nodes 400, 600 and 800 respectively. The average degree of nodes in these graphs is 11. The average channel balance is 7 and the average value of PBTs is .5. We simulate the PMTs and balanced routing algorithm in R. In each experiment, we execute 1000 PBTs. We compare the performance of our routing algorithm with a routing procedure that always uses the shortest path from the sender to the receiver. Majority of routing algorithms for PBT aims to develop a fast routing algorithm. Hence we use the shortest paths between the parties as to the outcome of such routing algorithms. We use three parameters to evaluate the performance of our routing algorithm as (a) standard deviation of channel values (high value of the standard deviation indicates that channels are highly imbalanced) (b) number of successful PBTs and (c) path length of PBTs (it indicates the cost transfer). The outcome of these experiments is shown in Fig. 3. First three plots show the standard deviation of the channels. It clearly shows that standard deviation of channels is increasing over the time for shortest path based routing and the same remains almost constant for our balanced routing algorithm. The last figure shows (a) number of completed PBTs for balanced

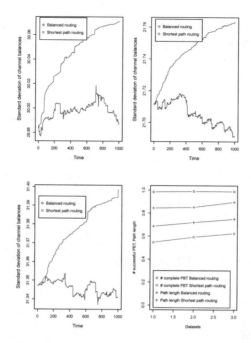

Fig. 3. Results: First three plots show the standard deviation of channels remains low for balanced routing. The last figure shows (a) number of completed PBTs for balanced routing is higher than the same for shortest path routing and (b) Average path length for PBTs is higher for balanced routing.

routing is higher than the same for shortest path routing and (b) Average path length for PBTs is higher for balanced routing. In the proposed routing protocol it is possible that a peer broadcasts wrong coordinates intentionally. But it is difficult to find appropriate coordinate to mislead the peers as by moving away from a set of peers, a peer may move closer to another set of peers. Also, we may use the reputation of peers to deter such behavior.

6   Conclusion

In this paper, we proposed a routing algorithm that can maintain a balanced channel network. Our routing algorithm preserves the privacy of the peers as they do not need to reveal exact channel balances. Using experimental evaluation we have shown that our algorithm maintains a balanced channel network and outperforms other routing algorithms which use the shortest path as the route for PBT.

Acknowledgement. This publication has emanated from research supported in part by a research grant from Science Foundation Ireland (SFI) and the Department of Agriculture, Food and the Marine on behalf of the Government of Ireland under Grant

Number SFI 16/RC/3835 (VistaMilk), co-funded by the European Regional Development Fund and the financial support of Science Foundation Ireland (SFI) under Grant Number SFI/12/RC/2289.

References

1. Raiden network. http://raiden.network/
2. Decker, C., Wattenhofer, R.: A fast and scalable payment network with bitcoin duplex micropayment channels. In: Pelc, A., Schwarzmann, A.A. (eds.) Stabilization, Safety, and Security of Distributed Systems, pp. 3–18. Springer, Cham (2015)
3. Malavolta, G., Moreno-Sanchez, P., Schneidewind, C., Kate, A., Maffei, M.: Anonymous multi-hop locks for blockchain scalability and interoperability. In: NDSS (2019)
4. Green, M., Miers, I.: Bolt: anonymous payment channels for decentralized currencies. In: Proceedings of the 2017 ACM SIGSAC Conference on Computer and Communications Security. CCS 2017, pp. 473–489. ACM, New York (2017). https://doi.org/10.1145/3133956.3134093
5. Grunspan, C., Pérez-Marco, R.: Ant routing algorithm for the lightning network CoRR abs/1807.00151 (2018). arXiv:abs/1807.00151
6. Khalil, R., Gervais, A.: Revive: rebalancing off-blockchain payment networks. In: Proceedings of the 2017 ACM SIGSAC Conference on Computer and Communications Security. CCS 2017, pp. 439–453. ACM, New York (2017). https://doi.org/10.1145/3133956.3134033
7. Malavolta, G., Moreno-Sanchez, P., Kate, A., Maffei, M.: Silentwhispers: enforcing security and privacy in decentralized credit networks. IACR Cryptology ePrint Archive 2016/1054 (2016)
8. Malavolta, G., Moreno-Sanchez, P., Kate, A., Maffei, M., Ravi, S.: Concurrency and privacy with payment-channel networks. In: Proceedings of the 2017 ACM SIGSAC Conference on Computer and Communications Security. CCS 2017, pp. 455–471. ACM, New York (2017). https://doi.org/10.1145/3133956.3134096
9. Poon, J., Dryja, T.: The Bitcoin Lightning Network:Scalable Off-Chain Instant Payments. https://lightning.network/lightning-network-paper.pdf
10. Prihodko, P., Zhigulin, S., Sahno, M., Ostrovskiy, A., Osuntokun, O.: Flare : an approach to routing in lightning network white paper (2016)
11. Reynolds, C.W.: Flocks, herds and schools: a distributed behavioral model. In: Proceedings of the 14th Annual Conference on Computer Graphics and Interactive Techniques. SIGGRAPH 1987, pp. 25–34. ACM, New York (1987). https://doi.org/10.1145/37401.37406

Anticipatory Policy as a Design Challenge: Experiments with Stakeholders Engagement in Blockchain and Distributed Ledger Technologies (BDLTs)

Denisa Reshef Kera[(⊠)] [iD]

BISITE, University of Salamanca, R + D + I Building, 37007 Salamanca, Spain
denisa.kera@usal.es

Abstract. We are proposing a combination of design and policy methods enabling anticipatory governance [4, 5, 7] of emerging blockchain and decentralized ledger technologies (BDLTs). Involving stakeholders in the early development and design of an emerging infrastructure is critical for the support of the Responsible Research and Innovation (RRI) agenda [8, 9, 14] and related calls for anticipatory governance. On the example of our Lithopia project [6] connecting satellite data with blockchain services, we will discuss the strategy of combining prototyping with future scenarios through a simulation game. We claim that this combination of policy deliberation and design supports anticipatory governance of blockchain and decentralized ledger technologies and engages the public in future-making.

Keywords: Anticipatory governance · Anticipatory design · Blockchain · Smart contracts

1 Introductionon

Users in most design methods are addressed as individuals that have various needs and desires (UX, human-centered design) supported by the right combination of technology and modalities [2, 3]. In the case of participatory and service design, users are addressed as part of an organization, community or another social group which poses further challenges [15]. Design has to respond and integrate various conflicting needs and desires of many users, their patterns in behavior, the context of the organizational culture, and the possibilities and limits of the existing technological infrastructure and regulations [11].

Lithopia project takes the challenges of participatory design, co-creation, and service design further, and views the users as stakeholders in the future. In this project, we explore the possibility of design as future-making [13] that reacts directly to the policy challenges involved in the emerging technologies, such as the blockchain. The user in the case of emerging technologies is a citizen and not only an individual or a group member. S/he is someone deeply invested in the possibility of anticipatory governance of such emerging technologies that create regulatory uncertainties, exaggerated expectations, but also fears. Design for anticipatory governance supports the capacities

© Springer Nature Switzerland AG 2020
J. Prieto et al. (Eds.): BLOCKCHAIN 2019, AISC 1010, pp. 87–92, 2020.
https://doi.org/10.1007/978-3-030-23813-1_11

of "foresight, engagement, and integration" as a way of "managing emerging knowledge-based technologies while such management is still possible" [5].

To support such early engagements with emerging infrastructure, the users in the Lithopia project are defined as citizens in a simulation of village that that uses functional prototypes of near future services connecting satellite data and the blockchain and distributed ledger technologies (BDLTs). The simulation and the prototypes train them to think as stakeholders in the future of this community shaped by the emerging new infrastructure. As future citizens, they face conflicting interests and expectations about the technology as much as about society, and the project provides them with means to connect deliberation with prototyping. To induce such experience of a future citizenship invested in prototyping and data governance in the case of the blockchain technologies they interact and tweak several functional prototypes in the simulation game of a "smart village" called Lithopia.

2 Prototyping in the "Smart Village" of Lithopia

The workshop uses templates of a Lithopia Hyperledger Fabric based contracts supporting transactions, such as become a Lithopian, offer a partnership and property, to reveal how blockchain services operate on the level of code and network communication. We enable the participants to prototype while deliberating on the life in Lithopia as stakeholders in ist future. The first activity is to make everyone a citizen of Lithopia through a smart contract experienced directly over a Node-RED interface that communicates with the REST API Hyperledger Fabric services installed on our server.

In the next step, they all see a design fiction movie that depicts a typical "sunny" day in Lithopia with villagers performing strange gestures to trigger social contracts, such as becoming a friend, partner, buying or selling a property. We confront the participants with a near future scenario of smart contracts triggered by satellite and drone data that act as notaries. Everyone is then given a template to deliberate upon it and even change or hack the contracts or even demand a moratorium on such technology. After participants become citizens of Lithopia, they have to propose a partnership or sell and buy property, and then follow how the data are saved on the ledger, and how they are triggered and changed by the smart contracts utilizing outside APIs and services.

In order to change ownership or partnership status in Lithopia, you have to be at the right place at the right time (GPS locations visible to Sentinel 2A and B satellites) and perform special gestures, move big LiCoins as objects or even cover a 10 × 10 m area visible as a pixel to the Copernicus satellites. While revealing the functional prototype and explaining the template (as well as the basic terminology or permissioned blockchain services, APIs etc.), we are slowly introducing the rich narrative of Lithopia and its special relation to the Micronesian island of Yap that uses large stone coins to preserve their oral memory of ownership, marriages, and important events.

Lithopians deploy their carpets or large 3D printed LiCoins visible to satellites and drones to trigger smart contracts, but they do not use any tokens or currency. In the plastic of these large LiCoins, Lithopians mix and hide the illegally obtained lithium from the old mines to reclaim the ownership of their natural resources. Their smart contracts are a form of oral culture timestamping that emphasizes genealogy over exchange and stewardship over ownership. The story also explores a form of resistance that is not direct, but creative and exploratory.

The project was inspired by the traditional mining region of Cínovec in the Czech Republic and its resistance to the interests of the mining industries trying to extract its rich lithium deposits. The prototypes and this story then set up a stage to discuss the current hype of national cryptocurrencies and other speculative investments in emerging technologies. Lithopia mocks the speculative ICOs, national cryptocurrencies, but also the political promises linked to Lithium reserves in the Czech Lands, and offers this as a model for other similar sites that pose the challenge of the commons.

3   Imagining Alternative Futures and Design Through Lithopia

Lithopia project combines storytelling and prototyping, deliberation and testing to give a direct experience with the design process, but also policy issues - automation and algorithmic governance, privacy and transparency in the age of satellite imagining, and the most critical question of who regulates the future infrastructure. The Lithopias gradually learn to deliberate upon these issues by using arguments as much as code.

Prototyping plays a central role in this exercise of anticipatory governance of BDLTs. It is used not only as a design method for gathering requirements or feedback, but also as a policy tool that provides strategic, future-oriented planning often addressed by various foresight and future scenario methods. After the Lithopias experience the tools (dashboard, REST API server) and activities in our simulation game, we ask them to tweak or design their own prototype, which they also present, and then everyone has to vote on the future of the blockchain in Lithopia.

Participants decide as a group whether Lithopia should continue or discontinue the use of blockchain based smart contracts with or without any outside supervision and regulation. We ask every participant to list three main reasons for/against, their expectations, fears, but also proposal for regulations or competencies and forms of supervision. In the final step, based on the voting, they finish the simulation and continue with collaborative two axes future scenario exercise on blockchain and governance to summarize the experience and define four possible scenarios for the future. In this part, the goal is to follow how the experience of prototyping and "living" in Lithopia influenced the final vote and the four possible futures.

4   Bridging Policy and Design Divides

With Lithopia project we want to bridge the divide between designing and policy-making to support anticipatory governance. The emerging technologies profoundly challenge not only our existing infrastructures but also regulations and governance and we need better methods to tackle this. Policy and regulations do not need to be just some ex-post strategies that come after a significant technological challenge and as a response to a crisis. This reactive view of policy is not efficient and just cements the distrust in the public institutions and infrastructures. Present calls for RRI and antici-patory governance are trying to change this reactive role of current governance and offer a more pro-active and even anticipatory practice that can prepare the society and different stakeholders to the new challenges, but they remain discoursive.

AnticipatoryLedgers project has a goal to define a framework for using design and future scenario methods to support anticipatory governance of BDLTs. It combines prototyping with testing and engaging stakeholders to not only define their needs, fears, and expectations but also to negotiate, adapt and envision desirable future for such future infrastructure.

In this paper, we described the process of defining the framework and the use case which we plan to apply to test such anticipatory prototyping. The central hypothesis is that the practices of prototyping can support policy-making as more inclusive and democratic activity that empowers the users to feel as stakeholders. We are still looking for ways how to involve the various users in the early stages of development and we are taking inspiration from the DIY and maker movements [1, 12] along participatory design. The creation of new products, services, and infrastructure should not be left only to developers if we want to avoid the mistakes of ex post regulations.

To make this convergence of technology and governance more inclusive, we need to involve the ethical and policy reflections and considerations directly and early in the prototyping phase of the BDLTs applications. The present fragmentation of BDLTs with many white papers, competing platforms and speculative investment related to future scenarios is an ideal stage to engage the citizens in the future making. Though prototyping, we can involve the public and the different stakeholders directly in the processes of making the future infrastructure together rather than imposing it by claiming better, faster and more secure algorithms.

The goal of the Lithopia project is to "prototype" ethical and governance frame-works and protocols simultaneously rather than in parallel. In this sense, our methods could help the BDLTs to move from their utopian sentiments of the early manifests of cyberspace and Internet, which detested governance and ethics external to code and technology, to more mature interest in different concepts and theories of ethics and management through prototypes. Instead of claiming that every new platform will make society more just and free through the work of one group - hackers and pro-grammers - we need cooperation that will pick up the problems and issues early by working on the prototypes. For the policy scholarship on emerging technologies, our project and proposal of working directly on the BDLTs application provide a new method to test the hypothesis and demonstrate the concept.

5 Summary

This experimental, design and policy-driven research tries to use productively the current tensions and convergences between emerging technology (BDLTs) and "ven-erable" issues in governance and ethics. The BDLTs as a crucial future infrastructure for governance shows a lack of direct engagements of the public and the different stakeholders in the processes of its development, testing, and implementation of the applications. Similarly, the various STS concepts, which advocate direct and early engagements with emerging technologies through interdisciplinary and interactive "socioethical engagements" [16, 17], "upstream engagements" [18, 19] or "technologies of humility" [20], lack a more design-oriented or prototyping focus, which defines any emerging technology and infrastructure. The proposed research agenda (AnticipatoryLedgers) and Lithopia project hope to bridge these divides.

The goal of Lithopia project is to use design and policy methods to support various stakeholders in the early phases of development rather than only in the adoption of emerging technologies. We use the anticipatory governance as a framework for prototyping that can capture the complex social, economic and political outcomes resulting in so-called "mediated (future) scenarios" [10].

The Lithopia prototypes of smart contracts serve to enable participants to evaluate neglected issues of BDLTs governance (accountability, shared responsibility, a division of powers), but also ethical principles (deontological versus utilitarian rules and laws) or philosophical questions (relation between hashes, concepts, codes, data). These prototypes and future scenarios help the stakeholders to define design and policy requirements for the BDLTs community, such as more hybrid governance models.

References

1. Nordmann, A.: Responsible innovation, the art and craft of anticipation. J. Responsible Innov. **1**, 87–98 (2014)
2. Guston, D.H.: Understanding 'anticipatory governance'. Soc. Stud. Sci. **44**, 218–242 (2014)
3. Davies, S.R., Selin, C.: Energy futures: five dilemmas of the practice of anticipatory governance. Environ. Commun. **6**, 119–136 (2012)
4. Reber, B.: RRI as the inheritor of deliberative democracy and the precautionary principle. J. Responsible Innov. **5**, 38–64 (2018)
5. Zimmer-Merkle, S., Fleischer, T.: Eclectic, random, intuitive? Technology assessment, RRI, and their use of history. J. Responsible Innov. **4**, 217–233 (2017)
6. Pellé, S.: Process, outcomes, virtues: the normative strategies of responsible research and innovation and the challenge of moral pluralism. J. Responsible Innov. **3**, 233–254 (2016)
7. Kera, D.: Github Lithopia contract. https://github.com/anonette/lithopia
8. David, B.: PACT: a framework for designing interactive systems. In: Designing Interactive Systems (2013)
9. Benyon, D.: Designing Interactive Systems: A Comprehensive Guide to HCI, UX and Interaction Design. Pearson/Education, Harlow (2013)
10. Routledge International Handbook of Participatory Design. Routledge (2013)
11. Stickdorn, M., Schneider, J.: This is Service Design Thinking: Basics, Tools, Cases. BIS (2011)
12. Imagined futures in science, technology and society. Routledge (2017)
13. Ames, M.G., et al.: Making cultures. In: Proceedings of the Extended Abstracts of the 32nd Annual ACM Conference on Human Factors in Computing Systems - CHI EA 2014, pp. 1087–1092. ACM Press (2014). https://doi.org/10.1145/2559206.2579405
14. Tanenbaum, J.G., Williams, A.M., Desjardins, A., Tanenbaum, K.: Democratizing technology. In: Proceedings of the SIGCHI Conference on Human Factors in Computing Systems - CHI 2013 2603. ACM Press (2013). https://doi.org/10.1145/2470654.2481360
15. Selin, C.: Merging art and design in foresight: making sense of Emerge. Futures **70**, 24–35 (2015)

16. Flear, M.L., Pickersgill, M.D.: Regulatory or regulating publics? The European union's regulation of emerging health technologies and citizen participation. Med. Law Rev. **21**, 39–70 (2013)
17. Akrich, M., Bijker, W., Law, J.: Shaping Technology/Building Society: Studies in Sociotechnical Change. MIT Press, Cambridge (1992)
18. Harvey, A., Salter, B.: Anticipatory governance: bioethical expertise for human/animal chimeras. Sci. Cult. (Lond) **21**, 291–313 (2012)
19. Rogers-Hayden, T.: Upstream engagement. In: Encyclopedia of Science and Technology Communication. SAGE Publications, Inc. https://doi.org/10.4135/9781412959216.n311
20. Jasanoff, S.: Technologies of humility: citizen participation in governing science. Minerva **41**, 223–244 (2003)

The Electronic Bill of Lading

Challenges of Paperless Trade

Stefan Wunderlich[(✉)] [iD] and David Saive [iD]

Carl von Ossietzky University of Oldenburg, Ammerländer Heerstraße 114-118,
26129 Oldenburg, Germany
{stefan.wunderlich,david.saive}@uol.de

Abstract. The bill of lading (B/L) is the most important document for sea freight transport. It proves that goods were taken over in the described form by a carrier. Moreover, it secures the obligation to deliver the goods to the place of destination and to deliver them to the consignee. In addition, the transfer of ownership of the good named in the B/L can be replaced by the transmission of the B/L. To this day, the B/L is still a paper-based document. Various attempts to digitalize the B/L failed due to heterogenous reasons. Especially approaches using a central intermediary have not been accepted by the market. This paper presents the domain-specific and legal context as well as a conceptual approach to digitalize the B/L. In this paper, domain-specific challenges for the digitalization of the B/L are described as well as a conceptual approach to overcome such challenges.

Keywords: Paperless trade · Bill of lading · Blockchain technology ·
Sea freight process

1 Introduction

Sea freight accounts for a large share of the global transport volume. In 2010, a total of 60,053 bio. ton-kilometers were covered. It is estimated that this figure will even quadruple by 2050 [1]. In Germany, over 316 million tons of goods were handled in the peak year of 2008, i.e. before the global economic crisis. In 2016, the economy was able to recover slightly, so that 292 million tons could be moved [2]. For all traded goods, freight documents, such as the B/L, are available in analogous form. The aim of this research is to replace these analogue bills of lading with digital, functionally equivalent tokens using blockchain technology.

Modern merchant shipping is characterized using a large number of different documents. The documentation obligations range from the keeping of ship certificates to the keeping of ship diaries and transport documents. Since 2018/01/01, these include: General Declaration (FAL Form 1), Cargo Declaration (FAL Form 2), Ship's Stores Declaration (FAL Form 3), Crew's Effects Declaration (FAL Form 4), and many more. In addition, all ships must carry various seaworthiness certificates, some of which are listed as examples: International Tonnage Certificate, International Load Line Certificate, Intact stability booklet, Damage control booklets, Minimum safe

© Springer Nature Switzerland AG 2020
J. Prieto et al. (Eds.): BLOCKCHAIN 2019, AISC 1010, pp. 93–100, 2020.
https://doi.org/10.1007/978-3-030-23813-1_12

manning document, etc. These documents are directly related to the cargo and must be traced [3]. The *bill of lading* (B/L) is still the most important document for freight transport. It fulfils several functions at once: On the one hand, it proves that the goods were taken over in the described form by the carrier. On the other hand, it secures the obligation to deliver the goods to the place of destination and to deliver them to the consignee. In addition, it serves as a negotiable document of title, which means that the transfer of ownership of the goods named in the B/L can be replaced by the transfer of the B/L [4].

2   Domain-Specific Background

In the context of overseas purchases, the B/L serves as basis for the letter of credit transaction between the banks of the buyer and seller. It becomes particularly clear how much the use of paper-supported B/L represents an anachronism in today's digitized world: So far it has been necessary for paper B/Ls to pass through the hands of the parties involved, i.e. from the seller to the seller's bank, then to the buyer's bank and finally into the hands of the buyer of the goods or the recipient of the goods, so that the parties involved can check the documents. Usually, many actors are involved in such logistics processes. Figure 1 shows that actors of sea freight processes are interconnected. This interconnection leads to error-prone and very slow processes because a considerable communication expenditure occurs.

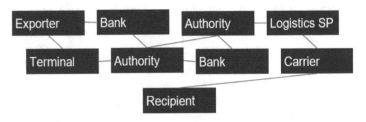

Fig. 1. Small excerpt of the actors involved in the logistics process

The procedure of document review is very time-consuming, which in the worst case takes longer than the actual transport. Therefore, it slows down the transport process and increases the price of the goods. In total, the transport-related documentation obligations alone account for 5–10% of the total costs of transport [5]. Against the background of the considerable cost pressure, it is therefore urgent to find a way of relieving the burden of paper-based sea freight documentation on the transport players. The German legislator has recognized the necessity of electronic B/L and introduced the new section 516 HGB (*Handelsgesetzbuch*, German Commercial Code) with the Act on the Reform of Maritime Trade Law of 2013 [3]. According to section 516 par. 2 HGB the paper-based B/L is equivalent to a digital representation, if it fulfils the same functions as the paper-based B/L [3]. The principle of functional equivalence thus enables electronic B/L in accordance with German law. This automatically raises the question of whether the industry has already been able to develop market-ready solutions to implement the economically necessary, politically desired and legally permissible electronic B/L. This is particularly interesting, because digitalization is the common goal of the maritime economy. Digitalization is indispensable in order to remain at the forefront of economic development.

Using electronic B/L, trading is enormously accelerated and ultimately made more cost-efficient. Particularly against the background of the still ongoing crisis in shipping, companies are dependent on further cost savings [6]. The complex paper-based documentation of the transport would be eliminated using electronic B/L, as would the complex manual document review processes.

In addition, governmental institutions, such as customs or import authorities, can also be included on the blockchain and therefore the logistics process. This would significantly speed up the import procedure by eliminating the need to check import documents manually. In addition to state control of the documents, there would also be no need for time-consuming document verification in the frame of the letter of credit business (uniform customs and practice for documentary credits/UCP 600) [7]. The banks would be integrated on the blockchain as actors with access and writing rights and would automatically validate the B/L data record. If the goods on the way are to be transferred, the acquirer can be included on the blockchain as a new node. Then the current owner of the token and thus of the B/L can send the token to him.

So far, several approaches have been developed to replace the paper form of the transport documents with digital equivalents [8]. The Bill of Lading Electronic Registry Organization (BOLERO) is the pioneer of this digitization effort [9]. It developed the technical environment as well as a legal framework construct (rulebook), which should enable the participants of the BOLERO system to use electronic B/L. The BOLERO standard could not assert itself because there was a lack of trust in the integration and reliability of the central server architecture of the system, the trading of BOLERO bills of lading was only possible between participants in the network and the rulebook used did not offer the necessary legal security [10].

Blockchain is a promising technology to avoid the conceptual mistakes of the BOLERO system. The distributed approach ensures that there is no single point of failure any more. The participants of the transportation processes do not need to trust in a single, centralized instance [11]. The consensus mechanism of the blockchain creates trust in the database itself. By replacing the paper-based B/L with a token, single ownership of the B/L is maintained throughout the whole process.

3 Implications for Legally Compliant Electronic B/L

Under technical aspects, a blockchain-based token seems to be the perfect solution to overcome paper-based B/L. But the best technical solution is worthless, if it cannot be used due to legal restraints. Therefore, it is indispensable to ensure compliance with all legal fields, that are touched by the technology.

Under German law the basis is set by section 516 par. 2 of the German Commercial Code (HGB) [3]. By this, the legal principle of functional equivalence is introduced. This principle derives from the failed try to unify all international rules concerning maritime trade in the *Rotterdam Rules* (United Nations Convention on Contracts for the International Carriage of Goods Wholly or Partly by Sea) [12]. It states, that every kind of electronic record of a B/L may be used, if it represents the functional equivalent of the paper-based B/L. In addition, authenticity and integrity of the record must be maintained at any time. This can be achieved using public key infrastructure. Germany is the only country in the world, that allows the usage of electronic B/L this simply. Therefore, the approach will strictly follow German law. This is possible using *choice of law clauses*. Choice of law clauses allow that for each transport process a certain legal location can be chosen. This legal location is the legal basis for the respective logistics process and thus for the electronic B/L [13].

The principle of functional equivalence both facilitates and impedes using electronic B/L at the same time. On the one hand, electronic records are now considered legally equivalent B/L, but on the other hand, if an electronic record does not comply with a certain legal field, it does not represent a functional equivalent, because it is illegal. Therefore, an application that implements electronic B/L must meet full functional equivalence to be legally compliant.

3.1 Identifying the Affected Areas of Law

Legally compliant electronic B/L are only possible, if all affected areas of law are identified (functional equivalence). Thus, an examination of the whole B/L process under legal aspects is necessary. In a first step, the basic functions of a B/L must be identified. In the means of German law, the B/L fulfills mainly four functions (section 513 ff. HGB).

- It is a receipt for the goods delivered,
- it works as evidence for the existence of a contract of carriage,
- it only authorizes the lawful holder to assert his rights under the contract of carriage,
- and it represents the goods themselves, being a document of title [3].

An electronic record must fulfill all these functions at the same time. These are only the regulations that are required by the laws governing the B/L and the maritime transport. Since the electronic B/L is an electronic record, all regulations concerning electronic legal communications must also be fulfilled. In the following, the most important are introduced.

- Under European and therefore German law it is necessary, to make sure, that the user of electronic legal communications can correct his input data, to prevent him from false agreements (sec. 312i of the German Civil Code) [14].
- If personal data is stored inside the database, the rules of the *General Data Protection Regulation (GDPR)* must be fulfilled. The *right to erasure* or better known as *right to be forgotten* from section 17 GDPR enables the data subject to force the controlling person to erase his or her personal data from the database [15]. If the database is used to distribute data between companies of the same market, antitrust law may be touched as well. Under sections 101 and 102 of the *Treaty of the Functioning of the EU (TFEU)* anti-competitive concerted practices are forbidden. Besides the listed examples therein, the exchange of market-relevant information may be considered as an antitrust violation [16].

These examples show, that the implementation of a legally compliant electronic B/L needs more than just the representation of the functions of an original paper-based B/L. It is necessary having legal expertise involved to identify the legal requirements. It goes without saying that the approach requires a lot more than software engineering. It requires a process which facilitates software engineering and legal expertise at the same time.

After the affected areas of law are identified, the requirements for a legally compliant electronic B/L must be derived from the legal analysis. The distributed character of the blockchain makes this process even more challenging. Particularly challenging is the task of making a blockchain fully GDPR-compliant because certain implications and requirements of the GDPR are contradictory to fundamental blockchain concepts.

One of the key features of every blockchain is, that once data is stored, it cannot be altered or deleted anymore. This irreversibility of the blockchain stands in direct contradiction to the *right to be forgotten* of section 17 GDPR [15]. Erasing data in the means of the GDPR requires the deletion of the data and not only the obliteration of the data. Thus, a fork of the blockchain would not be sufficient, because it does not delete stored data from the users' databases but creates a new branch. The particular use case of electronic B/L would not be GDPR-compliant if certain information cannot be deleted permanently. Conventional blockchain implementations are therefore not sufficient to meet legal requirements. In fact, only new approaches, such as redactable blockchains implement functionality such as permanent data deletion [17].

The consensus mechanism constitutes, that the users of the blockchain are not only distributing and sharing data inside the network but deciding which information shall be stored. Therefore, an analysis of all data stored inside is necessary, to make sure, that no antitrust-relevant information is distributed amongst the parties. This must be assured technically, so the algorithm prevents such data being stored [16].

The main challenge of the implementation of blockchain technology for electronic B/L therefore is to develop an approach, that integrates the advantages of blockchain technology, such as distributed consensus, shared data, and data security as well as legal obligations, that at the same time seem to contradict blockchain principles.

4 Conceptual Approach

The main technical challenges are the mapping of properties of physical titles to digital titles as well as the digitization and technical safeguarding of legal properties of the process. The essential technical characteristics are the secure, unambiguous connection of title to ownership, the uniqueness of the title of ownership and the process of transferring the title of ownership, the possibility of reversing an incomplete transfer of ownership. In the physical title of ownership, the link to the physical artifact is a description of the artifact that has been assessed as sufficiently accurate by all parties involved. The most striking difference between digital and physical artifacts is their ability to be copied. While physical objects can only be copied with - comparatively - great effort and never perfectly, digital artefacts can be copied as often and exactly as desired. It is very difficult to distinguish between copy and original. Therefore, a digital artefact does not possess uniqueness per se. In order to transfer a title of ownership, however, it must be ensured that the title of ownership exists only once in valid form. In the digital world, this can be achieved by a central instance managing the transfer of the digital artifact and thus defining a unique ownership status.

The essential function of blockchain technologies is to prevent so-called "double-spending", i.e. to ensure the uniqueness of mostly monetary artefacts [18]. Because this can be implemented in a distributed system without mutual strong trust assumptions - so-called "trustless" relationships - blockchain technologies are an interesting and promising solution for this application. In principle, blockchain technologies prevent transactions from becoming invalid post hoc. It is precisely the goal of these technologies to achieve a reliable, globally coordinated state. Nevertheless, it is necessary from a legal perspective to carry out special reverse transactions (due to contestation, revocation or withdrawal) that define a subsequent state in which certain transactions are revoked. To do this, the transaction language must be able to define such processes algorithmically. Subsequent changes can be implemented through novel approaches such as chameleon hashes [17].

The blockchain implements a global layer of trust for all actors of the sea freight process. For each sea freight process that involves an electronic B/L, a new instance of a blockchain is set up. Each actor in the process functions as a single node and holds the relevant data locally. Thereby, only the actors concerned gain visibility on the data. Due to the fully digital approach no media breaks occur. Thus, the data stored in the blockchain is not prone to errors. Subsequent changes on data stored can only be achieved through the implemented consensus algorithm. Figure 2 shows that the interaction between actors is facilitated through the blockchain layer. This layer gives actors simultaneous read access without the possibility to single-handedly manipulate data. The electronic B/L is directly integrated in each actors' business processes. The electronic B/L is an integral component of the systems landscape and therefore offers a similarity of access.

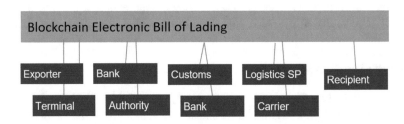

Fig. 2. The actors of the logistics process use a distributed ledger as shared database

Hence, a blockchain as a service concept arises where all actors of sea freight processes can participate on equal terms. *Blockchain as a service* offers companies in the logistics sector and in international trade the opportunity to participate in blockchain activities without developing their own solutions. Companies are considering offering blockchain as service solutions themselves, i.e. changing their own business model. In this way, logistics service providers would no longer only transport goods, but also data or intangible assets - possibly those related to physical transport.

The application of blockchain to implement electronic B/L is exposed to several risks, that must be countered by legal support. In general, the question must be answered how to deal with system errors regardless of their cause. System malfunctions can be caused by programming errors, power failures or malicious attacks. In order to ensure insurability despite all the care taken in incorrect transactions, limitation of liability should therefore be included in the terms of use, as is customary in transport law. However, the probability that faulty transactions will be executed is low. Due to the distributed network structure and the consensus mechanism, there is no longer a single point of failure. This also minimizes the risk of unauthorized access. In addition, the right to cancellation in the event of incorrect transactions must be included in the terms of use so that, in the event of reverse transactions, the right to reverse a transaction can be executed. The risk of a reversal of a transaction can be classified as high. There are many legal institutions that provide for the rescission of contractual relationships, for example following a rescission or incorrect legal transactions. In addition, comprehensive data protection transparency must be established so that the data subjects understand the legal basis of the processing, give their effective consent if necessary and are informed of their rights. The risk of an erroneous data protection agreement is also noteworthy because of the new GDPR and the associated legal uncertainty.

5 Outlook

Electronic B/L can be created, transferred and checked for correctness within very short time periods because paper-based processes and therefore media breaks are eliminated. The entire document transfer is thus considerably accelerated and much more resilient. The problem of delivery without submission of B/L can also be solved this way [19]. A blockchain-based approach on the problem described above enables the creation and trading of digital tokens, which in turn represent real values. These tokens will function as electronic B/Ls. However, the platform itself is to be created application-neutral in order to serve as a standard for any form of token-based trading. The permanent legal accompaniment of the development work ensures that the software always meets the legal requirements.

The cost- and time-intensive use of paper-based transport documentation is diametrically opposed to functioning and efficient transport sector. If it is possible to completely digitize the documentation of the transport processes, cargo means, such as containers, can be put back into circulation much earlier [20]. This also shortens the time ships spend in port, as the acceptance of documents can also be carried out completely digitally and in a matter of seconds. Ultimately, the working conditions on board are also improved and the crew can concentrate fully on the safety and smoothness of the traffic.

The weaknesses of the BOLERO approach (see Sect. 2) are prefaced by the fact that electronic B/L can only achieve a global breakthrough if low-threshold and low-cost access is guaranteed internationally and independently. Therefore, not only technical aspects are to be considered to implement electronic B/L but also operational and managerial ones. The handling of the electronic B/L should be entrusted to a an internationally operating, non-commercial organization. Under the umbrella of this independent organization, the private sector players in the international sea freight business become involved, as well as international organizations (e.g. IMO), and national and supranational authorities to ensure market acceptance throughout the maritime domain.

References

1. United Nations Conference of Trade and Development: Handbook of Statistics 2017. United Nations, Geneva (2017)
2. Destatis: Verkehr aktuell. Fachserie **8**, 1(1), Wiesbaden, Statistisches Bundesamt (2017)
3. Federal Republic of Germany: German Commercial Code (1900)
4. Gaskell, N.: Bills of Lading 2e: Law and Contracts. Routledge, Abingdon (2017)
5. United Nations. https://www.unece.org/fileadmin/DAM/trade/workshop/wks_capbld/unedocs_summary.pdf. Accessed 06 Feb 2019
6. Takahashi, K.: Blockchain technology and electronic bills of lading. J. Int. Marit. Law Publ. Lawtext Publ. Limited **22**, 202–211 (2016)
7. Debattista, C.: The new UCP 600: changes to the tender of the seller's shipping documents under letters of credit. J. Bus. Law **4**, 329–354 (2007)
8. Dubovec, M.: The problems and possibilities for using electronic bills of lading as collateral. Ariz. J. Int. Comp. Law **23**(2), 437–466 (2006)
9. Ma, W.: Lading without bills - how good is the Bolero bill of lading in Australia. Bond L. Rev. **12**, i (2000)
10. Takahashi, K.: Blockchain technology and electronic bills of lading. J. Int. Marit. Law **22**(3), 202–211 (2016)
11. Saive, D.: Blockchain documents of title – negotiable electronic bills of lading under German law. https://ssrn.com/abstract=3321368. Accessed 11 Feb 2019
12. Baatz, Y., Debattista, C., Lorenzon, F., Serdy, A., Staniland, H., Tsimplis, M.N.: The Rotterdam Rules: A Practical Annotation. CRC Press, Boca Raton (2013)
13. European Union: Regulation (EC) No. 593/2008 of the European Parliament and of the Council on the Law Applicable to Contractual Obligations (Rome I-Regulation) (2008)
14. Federal Republic of Germany: German Civil Code (1900)
15. European Union: Regulation (EU) 2016/679 of the European Parliament and of the Council of 27 April 2016 on the protection of natural persons with regard to the processing of personal data and on the free movement of such data, and repealing Directive 95/46/EC (General Data Protection Regulation; GDPR) (2016)
16. Louven, S., Saive, D.: Antitrust by design, the prohibition of anti-competitive coordination and the consensus mechanism of the blockchain. https://dx.doi.org/10.2139/ssrn.3259142. Accessed 11 Feb 2019
17. Ateniese, G., Magri, B., Venturi, D., Andrade, E.: Redactable blockchains – rewriting history in Bitcoin and friends. In: 2017 IEEE European Symposium on Security and Privacy (EuroS&P), pp. 111–126. IEEE, Paris (2017)
18. Pilkington, M.: 11 Blockchain technology: principles and applications. In: Research Handbook on Digital Transformations, vol. 225 (2016)
19. Debattista, C.: Bills of Lading in Export Trade. Bloomsburg Publishing, London (2008)
20. Goldby, M.: Electronic bills of lading and central registries: what is holding back progress? Inf. Commun. Technol. Law **17**(2), 125–149 (2008)

A Methodology for a Probabilistic Security Analysis of Sharding-Based Blockchain Protocols

Abdelatif Hafid[1]([⊠]), Abdelhakim Senhaji Hafid[2], and Mustapha Samih[1]

[1] Department of Mathematics, Faculty of Sciences, University Moulay Ismail,
B.P. 11201 Zitoune, Meknes, Morocco
abdelatif.hafid@yahoo.com, samih.mustapha@yahoo.com
[2] Department of Computer Science and Operations Research, University of Montreal,
Montreal, Canada
ahafid@iro.umontreal.ca

Abstract. In the context of blockchain protocols, each node stores the entire state of the network and processes all transactions. This ensures high security but limits scalability. Sharding is one of the most promising solutions to scale blockchain. In this paper, we analyze the security of three Sharding-based protocols using tail inequalities. The key contribution of our paper is to upper bound the failure probability for one committee and so for each epoch using tail inequalities for sums of bounded hypergeometric and binomial distributions. Two tail inequalities are used: Hoeffding and Chvátal. The first tail (Hoeffding inequality) is much more precise bound. The second (Chvátal inequality) is an exponential bound; it is simple to compute but weaker bound compared to Hoeffding. Our contribution is an alternative solution when the failure probability simulations are impractical. To show the effectiveness of our analysis, we perform simulations of the exponential bound.

Keywords: Blockchain · Sharding · Failure probability ·
Tail inequality · Hypergeometric distribution ·
Probabilistic security analysis · Exponential bound

1 Introduction

Blockchain is a technology that, when used, can have a great impact in almost all industry segments including banking, healthcare, supply chain and government sector. It can be simply defined as a distributed digital ledger that keeps track of all the transactions (e.g. asset transfer, storage) that have taken place in a secure, chronological and immutable way using peer-to-peer networking technology. It does not rely on any trusted central entity (e.g. bank) to validate transactions and extend the blockchain; the network nodes (aka miners), using a consensus protocol, agree on which node can create (i.e. mine) a valid block and append it to the blockchain. For example, when Proof-of-work consensus protocol [1]

© Springer Nature Switzerland AG 2020
J. Prieto et al. (Eds.): BLOCKCHAIN 2019, AISC 1010, pp. 101–109, 2020.
https://doi.org/10.1007/978-3-030-23813-1_13

is used, the node that first solves a mathematical puzzle, adds the block to the blockchain and gets rewarded (by the network and transaction fees). More specifically, a transaction is broadcasted to all the nodes in the network (1000 in the case of bitcoin); a node that receives the transaction, it checks whether the transaction is valid; if the response is yes, it sends the transaction to its neighbors; otherwise, it drops the transaction. Periodically (e.g. each 10 minutes in Bitcoin [1]), a block (includes a list of transactions; e.g., up to 4000 transactions in Bitcoin) is created/mined and broadcasted to all the nodes in the network; the node who mined the block (first to solve the mathematical puzzle), appends the block to the blockchain and broadcasts it to its neighbors. A node that receives a block, it validates the block; if valid, it appends the block to the blockchain and broadcasts to its neighbors; otherwise, it drops it. Thus, in general, all nodes have the same copy of the blockchain; if not, nodes builds on the longest chain. One of the key limitations of proof-of-work based blockchains is scalability; indeed, the number of transactions that can be processed per second is small (e.g. up to 7 for Bitcoin and 15 for Ethereum [2]). This is unacceptable for most payment applications that require 1000s of transactions per second (e.g. Visa and PayPal). The objective of blockchain scalability is to process a high number of transactions per second (i.e. throughput) without sacrificing security and decentralization [3,4]. Indeed, we can easily considerably increase the throughput but we will lose in terms of decentralization (wich is a key characteristic of blockchain).

A number of solutions to scale blockchain have been proposed; we can classify them into two categories: (1) On-chain solutions: they propose modifications to the blockchain protocols, such as Sharding (e.g. [5–7]) and block size increase (e.g. [8]); and (2) off-chain solutions (aka layer 2 solutions): these are built on the blockchain protocols; they process certain transactions (e.g. micro-payment transactions) outside the blockchain and only record important transactions (e.g. final balances) on the block- chain. Examples of layer 2 solutions include Lightning Network [9], Raiden Network [10], Plasma [11], and Atomic-swap [12]. Security and decentralization should be taken into account while solving the scalability issue in public blockchains. This is called the scalability trilemma; indeed, finding a balance between scalability, security and decentralization is very challenging. In this paper, we focus on analyzing the security of scalability solutions that use the concept of Sharding; this is motivated by the fact that Sharding is one of the promising solutions to the scalability problem. The basic idea behind Sharding is to divide the network into subsets, called shards; each shard will be working on different set of transactions rather than the entire network processing the same transactions. Several Sharding protocols have been proposed in the literature; they include Elastico [13], OmniLedger [14], RapidChain [15], Zilliga [16] and PolyChard [17]. Generally, Sharding is used in non-byzantine settings (e.g. [18]); Elastico [13] is the first Sharding-based protocol with the presence of byzantine adversaries. Elastico, divides the network into multiple committees where each committee handles a separate set of transactions, called shard. The number of shards grows nearly linearly with the size of the network. When the network grows up to 1,600 nodes, Elastico succeeds at increasing the throughput (e.g. up to 40 transactions per second (tx/sec)). However, it has shortcomings that include: (1) the randomness used in each epoch (i.e in each fixed time period; e.g., once a week) of Elastico can be biased by malicious nodes; and (2) it can only tolerate up to 25% of malicious/faulty nodes (total resiliency) and 33% of malicious nodes in each committee (committee resiliency). OmniLedger [14] has been proposed to fix some of the shortcomings of Elastico. In particular, it uses a bias-resistant public-randomness protocol to ensure security. The OmniLedger consensus protocol uses a variant of ByzCoin [19] to handle and achieve faster transactions (e.g. up 500 tx/sec when the network grows up to 1,800 nodes). Omniledger, like Elastico claims the same resiliency for both; total resiliency and committee resiliency. Recently, Zamani and Movahedi in [15] proposed RapidChain as a Sharding-based public blockchain protocol which succeeds at outperforming existing Sharding algorithms (e.g. [13,14]) in terms of scalability and security. Indeed, RapidChain can tolerate up to 33% of malicious/faulty nodes and 50% of malicious nodes in each committee. RapidChain claims a high throughput (e.g. up to 4,220 tx/sec when the network grows up to 1,800 nodes). Table 1 shows common characteristic of Sharding-based protocols used in our analysis. In this paper, we present a probabilistic security analysis of Elastico, OmniLedger and RapidChain. More specifically, we propose a probabilistic security analysis of these protocols using hypergeometric and binomial distributions. First, we calculate the failure probability for one committee; then, we calculate the union bound (i.e. the failure probability of each epoch); finally, we bound the failure probability with two bounds making use of the tail inequalities bounds [20–22]. The first bound [22] is much more precise tail bound ; the second [21] is an exponential bound which is more simple and elegant bound, however weaker bound compared to [22]. Thereafter, we upper bound the failure probability for each epoch by multiplying the committees bounds by the number of committees.

Table 1. Resiliency bound

Protocols	Total resiliency	Committee resiliency
Elastico [13] and Omniledger [14]	$\frac{1}{4}$	$\frac{1}{3}$
Rapidchain [15]	$\frac{1}{3}$	$\frac{1}{2}$

The contribution of this paper consists of a solution to analyze security (i.e. computing failure probability bounds) when failure probability simulation is unpractical (e.g. required number of simulations increases as the number of shards increases). To the best of our knowledge, this is the first time that Hoeffding [22] and chvátal [21] inequalities are used to analyze security of blockchain protocols. We implemented the exponential bound function [21] in order to verify and show the effectiveness of our analysis.

The paper is organized as follows. Section 2 presents the proposed probabilistic analytical model. Section 3 evaluates the model. Finally, Sect. 4 concludes the paper.

2   Analytical Model

2.1   Notations

Table 2 shows the notations we used in the paper. Note that the cumulative hypergeometric distribution $H(K, N, n, k)$ is the sum for all $i \geq k$ of the probability distribution function $h(K, N, n, i)$.

Table 2. Notations

Notation	Meaning
N	The total number of nodes
n	The committee size
K	The total number of malicious nodes
n_c	The number of committees
p_c	The committee failure probability
p_0	The bootstrap probability for RapidChain
$h(K, N, n, k)$	The hypergeometric distribution with parameters K, N and n
$H(K, N, n, k)$	The cumulative hypergeometric distribution with parameters K, N and n
$B(n, p, k)$	The cumulative binomial distribution with parameters n and p
X	Random variable which represents the number of malicious nodes

2.2   Probability Distributions

We use the hypergeometric and binomial distribution to calculate the failure probability for one committee and then for each epoch. We define the probability that a committee contains k malicious nodes sampled from a population of N nodes containing at most K corrupt nodes. Let X denote the random variable corresponding to the number of malicious nodes in the sampled committee. If we assume that X follows the hypergeometric distribution with parameters K, N and n, the failure probability is:

$$h(K, N, n, k) = \frac{\binom{K}{k}\binom{N-K}{n-k}}{\binom{N}{n}} \tag{1}$$

In this paper, we are interested in the probability that there is X, smaller than k malicious nodes when randomly selecting a committee of n nodes without replacement from a population of N nodes containing at most K corrupt nodes.

The cumulative hypergeometric distribution function allows us to calculate this failure probability; indeed, the failure probability for one committee for Elastico and OmniLedger is:

$$H(K, N, n, \frac{n}{3}) = \sum_{k=\lfloor \frac{n}{3} \rfloor}^{n} \frac{\binom{K}{k} \binom{N-K}{n-k}}{\binom{N}{n}} \tag{2}$$

In general, when the hypergeometric distribution is used, a comparison is performed with the binomial distribution. More specifically, it is said that if n is small relative to the population size N, then X could be approximated by a binomial distribution. Practically, we approximate hypergeometric distribution by a binomial distribution when the sample size is smaller than 10% of the population [23]. However, when the sample size gets larger relative to the population size, it is recommended to use the hypergeometric distribution (the hypergeometric distribution yields a better approximation in this case). If the sampling is done with replacement, we use the cumulative geometric distribution [20] or cumulative binomial distribution [23] instead of the cumulative hypegeometric distribution to calculate the failure probability. Now, if we assume that $X \sim B(n, p)$ (i.e. X follows the binomial distribution with parameters n and p) where $p = \frac{K}{N}$, p is the probability of each node being malicious. Thus, the failure probability of one committee for Elastico and OmniLedger using the cumulative binomial distribution function is:

$$P(X \geq \frac{n}{3}) = \sum_{k=\lfloor \frac{n}{3} \rfloor}^{n} \binom{n}{k} p^k (1-p)^{n-k}. \tag{3}$$

2.3 Tail Inequalities

The main contribution of our work is to upper bound the failure probability for one committee and so for one epoch using two bounds functions. The tail inequalities are powerful results that can be compute these bounds [20–22]. Firstly, we upper bound the failure probability for one committee as well as for each epoch. The following bound is given by Hoeffding [22]:

$$H(K, N, n, k) \leq G(x), \tag{4}$$

where

$$G(x) = \left(\left(\frac{p}{p+x} \right)^{p+x} \left(\frac{1-p}{1-p-x} \right)^{1-p-x} \right)^n, \tag{5}$$

$p = \frac{K}{N}$ and $k = (p+x)n$ with $x \geq 0$.

Hence, we can bound the failure probability of one committee for Elastico and OmniLedger as follows:

$$H(K, N, n, \frac{n}{3}) \leq G(x), \tag{6}$$

where

$$x = \frac{1}{3} - p, \qquad (p \leq \frac{1}{4}).$$

Likewise, we upper bound the failure probability of one committee for Rapid-Chain:

$$H(K, N, n, \frac{n}{2}) \leq G(x), \tag{7}$$

where

$$x = \frac{1}{2} - p, \qquad (p \leq \frac{1}{3}).$$

The binomial distribution coincidentally has an analogous tail bound [21], which means:

$$B(n, p, k) \leq G(x), \tag{8}$$

where

$$B(n, p, \frac{n}{2}) = \sum_{k=\lfloor \frac{n}{2} \rfloor}^{n} \binom{n}{k} p^k (1 - p)^{n-k}.$$

Now, we upper bound the failure probability of each epoch for RapidChain; we calculate the union bound over n_c committees, where each committee can fail with probability p_c. When the sample size is smaller than 10%, p_c is calculated using cumulative binomial distribution. Otherwise, we use the cumulative hyper-geometric distribution. In the first epoch for RapidChain protocol, the committee election procedure can fail with probability $p_0 = 2^{-26.36}$ (see [15]). Thus, the failure probability for one epoch for RapidChain is upper bounded as follows:

$$p_0 + n_c p_c \leq V(x), \tag{9}$$

where

$$V(x) = p_0 + n_c G(x), \qquad n_c = \frac{N}{n}.$$

Secondly, Chvátal [21] propose another tail bound, it is simple and elegant (i.e. exponential function), but weaker bound compared to the last one. We obtain the following bound:

$$H(K, N, n, k) \leq F(x), \tag{10}$$

where

$$F(x) = \exp^{-2x^2 n}.$$

Thus, the failure probability for one epoch for RapidChain is bounded as follows:

$$p_0 + n_c p_c \leq U(x), \tag{11}$$

where

$$U(x) = p_0 + n_c F(x), \qquad n_c = \frac{N}{n}.$$

Similarly, we can upper bound the failure probability for each epoch for Elastico and OmniLedger.

3   Results and Analysis

Figure 1 shows the exponential tail bound and the probability of failure calculated using the hypergeometric and binomial distributions to sample a committee without replacement with various sizes from a pool of 2,000 nodes. In particular, Fig. 1(a) shows the plot of the failure probability for one committee as well as the exponential function bound in RapidChain. We observe that the exponential bound curve looks similar to the curve of the failure probability calculated when the committee size increases (when it approaches 100). Hence, we get a good approximation bound when the committee size gets larger. Figure 1(b) shows the plot of the exponential tail bound of the failure probability for one committee in the OmniLedger and Elastico and the failure probability both decrease when the committee size increases; in addition, when the committee size increases above 250 nodes, both curves look similar. Finally, Fig. 1(c) presents the shows of the exponential bound of the failure probability for one epoch in RapidChain and the failure probability for the union bound over the number of committees while varying the committee size. We conclude that our proposal allows to compute bounds with good precision especially in the case of larger committee sizes.

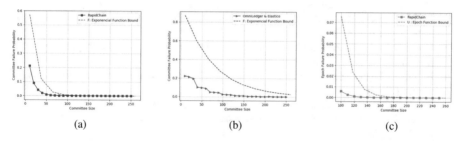

(a)   (b)   (c)

Fig. 1. Plot of the exponential bounds, as well as the failure probability vs. the committee sizes; (a) for one committee for RapidChain [15], (b) for one committee for Elastico [13] and OmniLedger [14], and (c) for one epoch for RapidChain [15].

4   Conclusion and Future Works

In summary, we proposed two bounds of the failure probability for one committee, thereafter for each epoch when we use the hypergeometric or the binomial distribution using tail inequalities. The first bound is more precise, but difficult to compute. The second is a simple exponential bound whereas weaker bound compared to the last one. We also calculated the failure probability for one committee as well as for one epoch using hypergeometric and binomial distributions. We have approximated the hypergeometric distribution with the binomial distribution when the sample size is smaller than 10%. We have implemented the exponential bound and the failure probability to show the performance of our analysis. We conclude that our proposal can be used to analyze security of any

Sharding-based protocol. For the future work, we will apply tail bounds which are more precise and can yield good approximations. Another interesting work is to make a probabilistic security analysis of Ethereum-Sharding.

References

1. Nakamoto, S.: Bitcoin: a peer-to-peer electronic cash system (2008)
2. Wood, G.: Ethereum: a secure decentralised generalised transaction ledger. Ethereum Proj. Yellow Pap. **151**, 1–32 (2014)
3. bloXroute Team: The scalability problem, (very) simply explained. https://medium.com/bloxroute/the-scalability-problem-very-simply-explained-5c0656f6e7e6. Accessed 28 Mar 2019
4. BitRewards: Blockchain scalability: the issues, and proposed solutions. https://medium.com/@bitrewards/blockchain-scalability-the-issues-and-proposed-solutions-2ec2c7ac98f0. Accessed 16 Mar 2019
5. Zhang, R.: Here's The Deal on Sharding. https://medium.com/coinmonks/heres-the-deal-on-sharding-96d8591856c4. Accessed 28 Mar 2019
6. Wang, H.-W.: Ethereum sharding: Overview and finality. https://medium.com/@icebearhww/ethereum-sharding-and-finality-65248951f649. Accessed 10 Mar 2019
7. Gencer, A.E., van Renesse, R., Sirer, E.G.: Short paper: Service-oriented sharding for blockchains. In: International Conference on Financial Cryptography and Data Security, pp. 393–401. Springer (2017)
8. Garzik, J.: Block size increase to 2MB. In: Bitcoin Improvement Proposal, vol. 102 (2015)
9. Poon, J., Dryja, T.: The bitcoin lightning network: scalable off-chain instant payments (2016)
10. Raiden Network Review: Fast, cheap, scalable token transfers for ethereum (2018)
11. Poon, J., Buterin, V.: Plasma: scalable autonomous smart contracts, White paper, pp. 1–47 (2017)
12. Komodo: Advanced blockchain technology, focused on freedom (2018)
13. Luu, L., Narayanan, V., Zheng, C., Baweja, K., Gilbert, S., Saxena, P.: "A secure sharding protocol for open blockchains. In: Proceedings of the 2016 ACM SIGSAC Conference on Computer and Communications Security, pp. 17–30. ACM (2016)
14. Kokoris-Kogias, E., Jovanovic, P., Gasser, L., Gailly, N., Syta, E., Ford, B.: OmniLedger: a secure, scale-out, decentralized ledger via sharding. In: 2018 IEEE Symposium on Security and Privacy (SP), pp. 583–598. IEEE (2018)
15. Zamani, M., Movahedi, M., Raykova, M.: Rapidchain: scaling blockchain via full sharding. In: Proceedings of the 2018 ACM SIGSAC Conference on Computer and Communications Security. ACM, pp. 931–948 (2018)
16. Team, Z., et al.: The zilliqa technical whitepaper (2017)
17. Li, S., Yu, M., Avestimehr, S., Kannan, S., Viswanath, P.: Polyshard: Coded sharding achieves linearly scaling efficiency and security simultaneously (2018). arXiv preprint arXiv:1809.10361
18. Danezis, G., Meiklejohn, S.: Centrally banked cryptocurrencies (2015). arXiv preprint arXiv:1505.06895
19. Kogias, E.K., Jovanovic, P., Gailly, N., Khoffi, I., Gasser, L., Ford, B.: Enhancing bitcoin security and performance with strong consistency via collective signing. In: 25th {USENIX} Security Symposium ({USENIX} Security 16), pp. 279–296 (2016)

20. Skala, M.: Hypergeometric tail inequalities: ending the insanity (2013). arXiv preprint arXiv:1311.5939
21. Chvátal, V.: The tail of the hypergeometric distribution. Discrete Math. **25**(3), 285–287 (1979)
22. Hoeffding, W.: Probability inequalities for sums of bounded random variables. In: The Collected Works of Wassily Hoeffding. pp. 409–426. Springer (1994)
23. Wroughton, J., Cole, T.: Distinguishing between binomial, hypergeometric and negative binomial distributions. J. Stat. Educ. **21**(1) (2013)

The "Tokenization" of the eParticipation in Public Governance: An Opportunity to Hack Democracy

Francisco Luis Benítez Martínez[✉][ID],
María Visitación Hurtado Torres[✉][ID],
and Esteban Romero Frías[✉][ID]

University of Granada, Granada, Spain
flbenitez@correo.ugr.es, {mhurtado,erf}@ugr.es

Abstract. Currently *Distributed Ledger Technologies*-DLTs, and especially the Blockchain technology, are an excellent opportunity for public institutions to transform the channels of citizen participation and reinvigorate democratic processes. These technologies permit the simplification of processes and make it possible to safely and securely manage the data stored in its records. This guarantees the transmission and public transparency of information, and thus leads to the development of a new citizen governance model by using technology such as a BaaS (Blockchain as a Service) platform. G-Cloud solutions would facilitate a faster deployment in the cities and provide scalability to foster the creation of Smart Citizens within the philosophy of Open Government. The development of an eParticipation model that can configure a tokenizable system of the actions and processes that citizens currently exercise in democratic environments is an opportunity to guarantee greater participation and thus manage more effective local democratic spaces. Therefore, a Blockchain solution in eDemocracy platforms is an exciting new opportunity to claim a new pattern of management amongst the agents that participate in the public sphere.

Keywords: Governance · Blockchain · eParticipation

1   Introduction

Since 2009 when Nakamoto [1] described the procedures to use Blockchain technology as a cryptographic management system in a single and distributed registry, he opened the way to transform economy but also a large part of our social and political system.

Thanks to this, the development of applications in the Fintech sector has continued to be deployed in the Internet with a set of dApps (mobile applications using blockchain technology that can be public or permissioned) that allow its multi-screen integration in a distributed manner. In parallel, there are new types of DLTs that progress in the development of new formats and possibilities due to more advanced node topologies such as Hedera Hashgraph [2] or Tangle, from the IOTA Foundation [3].

Developing a DLT is an opportunity to manage complex processes that simplify governance practices. They can guarantee the structure and security of the data. They

© Springer Nature Switzerland AG 2020
J. Prieto et al. (Eds.): BLOCKCHAIN 2019, AISC 1010, pp. 110–117, 2020.
https://doi.org/10.1007/978-3-030-23813-1_14

are an opportunity for the management of transparency and transactions that take place in public administrations. Another important feature is the capacity to temporally seal and thus guarantee the inalterability of the data contained. It allows the configuration of effective management systems [4] to achieve better accountability, through standard and scalable eGovernment models at all territorial levels of the administration.

The digital transformation of the administration in the service of citizen empowerment is a governmental obligation and a civic demand that is possible and necessary, not only because of the disruption that the DLTs suppose, but also the Cloud solutions and the possibility of managing the tools on platforms. There are methodologies for this, such as the one proposed by Huang and Karduck [5] and we have the opportunity to experiment with new solutions for democratic processes that allow new management systems for public governance.

Therefore, Blockchain technology is a unique opportunity to manage new governance processes in the environments of public institutions for the improvement of their efficiency and effectiveness [6] especially if it is used as a platform [7]. And go beyond e-voting solutions as Estonia has experimented in the last years in order to enhance citizen participation [8].

This paper proposes a new model of eParticipation system that integrates the possibilities that DLT systems [9] provide in relation to public governance processes and new methods of developing new tools for citizen empowerment To this end, we will describe a "tokenizable" participation system that helps local administrations to improve participatory projects using electronic voting tools based on blockchain. These tools are an effective means of improving citizen participation in the management of the everyday administration of cities.

The use of a G-Cloud system will be proposed to allow its optimal deployment with a high capacity of scalability and reproducibility in any environment, following a proven methodology for it [10].

This paper proposes the application of new processes and procedures that use blockchain as a platform within a G-Cloud based framework. The objective is to provide citizens with the opportunity to actively engage in local eParticipation projects. The structure of the paper is as follows. The next section presents a literature review of successful platforms used in the study in order to contextualize the proposed model. The third section describes this model and underlines the evident need of new ways of governance within the Smart City context in the form of a tokenizable eParticipation model. The final section highlights the usefulness of this innovative eParticipation model using blockchain as a platform to engage citizen participation and gives directions for future work.

2   Technological Background and Related Works

Currently there is a series of platforms and tools developed that can be the technological basis for the development of different governance projects in institutions and companies. Below we describe the most relevant:

1. Bitnation [11] is a decentralized platform for Governance 2.0. It is a political experiment from a voluntary, transnational and borderless perspective, which explores new democratic and social processes using the Blockchain technology. This platform was deployed in July 2014. In it, the first marriage using this technology was managed, alien to traditional institutions. Also, new systems of personal identification outside the control of the state and legal systems with digital notaries within it are experimented with. For the development of these social innovations, the Pangea project is used, which is a decentralized market for legal services, where users can experience new services of citizen governance. All their Smart Contracts are implemented in Ethereum.

2. Follow My Vote [12] is designed as a platform to guarantee electronic voting with high security, allowing a high degree of transparency in the democratic processes in which it is used. It can also be developed without compromising the privacy of the voters who use it, guaranteeing the traceability of their vote throughout the process until the closing of the voting and the final results. It is developed in open code, so anyone can audit the code used. It has been deployed both in UN votes and at a less rigid level, such as the television show "The Voice" in its North American version.

3. BoardRoom [13] is a governance management system that integrates both a dashboard and a dApp, which can be used both individually and by companies to manage their Smart Contracts, equally in public and permissioned blockchains developed in Ethereum. It serves both to issue tokens in crowdfunding projects, proxy voting systems for shareholders in Committees or Shareholders' Meetings and the development of payment systems through tokens in industrial consortiums.

4. Consul [14] is the most developed citizen participation tool to date for open government projects in municipal environments. It is developed in free software (its code is complete in GitHub) and has established the DECIDE platform to manage the eParticipation systems that it integrates. It is in use in 33 countries, and is used by 100 institutions such as the municipalities of Madrid (Spain), Paris (France) and Buenos Aires (Argentina). It has a main advantage, in that it has allowed the development of a scalable and highly tested system in several institutions around the world, but despite a high degree of reliability it can not guarantee the level of privacy and transparency that would allow it to be used as a standard.

5. TIVI [15] is the platform developed by Smartic and Cybernetica for verifiable voting using biometric factors from the voting point. Similarly to Consul, it uses a blockchain based ballot box and also uses encryption systems to secure the votes. They have developed voter authentication systems in Argentina and Armenia in 2017 and their first inclusion in systems of choice with Blockchain technology was in 2016 for the Primary elections of the Republican Party of Utah in the United States. Their actions are more oriented to electoral automatization processes than to the development of eParticipation systems. TIVI currently supports the *i-voting* Estonian system. [https://e-estonia.com/solutions/e-governance/i-voting/]

3   New Approach for a "Tokenizable" e-Participation Model

3.1   The Governance Deficit

Currently the concept of governance is a word devoid of meaning. Society does not perceive a risk management system and the uncertainty generated by globalization. Within a system where profits are minimized and losses are maximized, it is critical to redefine what good governance consists of.

It is necessary to establish a new governance system that develops a participatory system of governance, taking advantage of the technological opportunities that we have today. And this concept allows civic and technological empowerment to citizenship.

Governance is, in short, the construction of a proactive citizen governance, that is, a citizen with the capacity and possibility of taking an active part in the daily government, thanks to the development of a new democratic model through the use of the tools that the eDemocracy provides.

3.2   The Role of Open Smart Cities and the "Smart Citizens": Opening Local Governments

Our approach is part of the open government systems that have a series of digital tools for the development of eDemocracy solutions in the ambit of public participation. Therefore, this approach requires reconsidering the relationship between technology, people and cities from a perspective centered on citizenship and with its complex and contradictory vision of the urban concept [16].

The philosophy of the Smart City, whose development has been possible from the point of view of technology and Big Data, has led to its decision-making being replaced by a technocratic decision system based on the dictatorship of isolated data, and not from the point of view of the development of consensual public policies. Therefore, in the same way that the Open Government concept has been developed to establish new participation and accountability systems, it is necessary to establish a new concept: the Open Smart City.

Open Smart City is a philosophy that implies empowering citizens, that is, putting cities at the service of their citizens, empowering, in turn, public officials to develop the policies that are defined and agreed upon by the parties. It is the paradigm of governability (the form of governing), which not only governance (the processes and protocols that allow governing), which requires a more complex and permanent effort.

This vision implies giving space to new dynamics that come from society as citizen science, digital activism, "artivism", the "maker" movement, the policy and social labs [17], the co-creation and self-development communities, digital citizen movements, social hackers, citizen innovation laboratories [18], etc.

The idea presented aims to "open" the data and the processes that generate cities to make them available to society, becoming a common practice in the service of people. It provides an opportunity to create more democratic co-decision spaces in the political arena, creating the figure of Open Smart Citizens, i.e. citizens who operate in the spaces described above and that offer a new opportunity to generate new public policies and new democratic spaces.

Data mining and big data allow the development of new tools and techniques for the management of the city and its social diversity, from the point of view of social innovation.

Our proposal aims to support the growing demand of citizens to participate actively in the decisions that affect cities and the evolution of data mining techniques and their visualization tools. These allow the availability of information in real time, giving the people the possibility of having immediate data to manage a personal responsibility in the creation and design of services that affect them directly [19]. These processes of co-creation and co-participation are in the fundamental principles of open government and are what allow the deployment of the elements that form the Open Smart Cities format.

3.3   A "Tokenizable" eParticipation Model

The model we propose (Fig. 1) is based on a public G-Cloud system that allows the integration of a blockchain platform as a service (BaaS). This guarantees the functional requirement of scalability necessary in any governmental and/or territorial environment. In our system, the voter would access the platform through a permissioned dApp that would connect with the interface and the voter database to proceed with its recognition and identify him/her in the process.

Then the System Administrator would activate the Smart Contract in the front end that would allow the citizen to participate in the process opened by the agency to register their participation and the meaning of their vote. In the validation node, the encryption of the data is performed with a one-way hash function SHA-256 so that the result of the voting cannot be reversed. Once the vote is encrypted, the block will be added to the blockchain within the BaaS architecture used, and in turn the block generates the "token" that will grant the voter a series of advantages and/or rights of use in the city.

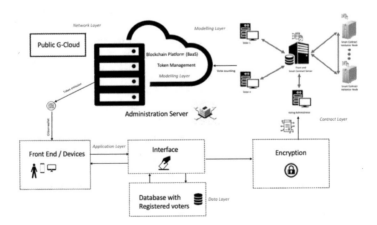

Fig. 1. Simplified representation of a blockchain e-voting participation

For the development of this proposal we have been taken into account previous studies of electronic voting that are being done in parallel by blockchain to solve different technical issues for its progressive implementation. Specifically:

a. Those that differentiate the use of blockchain for the development of voting systems using crypto currencies [20], such as those that use the Zcash protocol [21],
b. Those that rely on the deployment of Smart Contracts [22] with the limitations of options and number of voters that this entails, as is the case of BoardRoom,
c. And those that manage the blockchain as a "ballot box" as Follow My Vote and TIVI do [23].

Also there have been studies that have developed electronic voting systems that establish taxonomies and requirements [24, 25], the conceptualization of how to use blockchain to develop voting systems [26] or how to develop voting systems that allow more control of the process and its decisions to voters [27].

Taking into account that for the development of fintech projects for cryptocurrencies there is a large repository of projects and academic articles on their implementation, it is an opportunity to take advantage of all that experience to use "tokens" in a practical way in eDemocracy environments.

The idea proposed is to use "tokens" to reward the participation and co-management of projects of public utility.

The procedures to which it can be applied are varied, from accumulating them to exchange them for discounts in taxes or municipal fees for the participation in voting, the possibility of exchanging them to obtain cultural or sports advantages (tickets, subscriptions, discounts) or for the demonstration of the development of civic values through participation in the deployment of public policies.

With this proposal, a whole new field of citizen management and work is opened, which will help to increase participation ratios in the public processes and votes that are proposed, as we have been working in the MediaLab of the University of Granada, as well as in projects with the City of Granada on social innovation and co-management of public policies. We believe that it is a possibility to manage new developments of social innovation based on the deployment of a platform that is conceived as such. This will increase citizen participation and especially of the young population, by making use of dApps in any digital format.

4 Conclusions and Future Work

Currently it is necessary to experiment with new methods that allow a greater commitment of citizens to their local institutions and the blockchain technology allows it, thanks to one of its representative elements: the token.

In this work, a model based on blockchain has been proposed for the management of local governance, as a more efficient and inclusive way that opens new paths for the local management of territories. Its deployment follows a BaaS approach, which ensures scalability in a G-Cloud solution environment. This approach enables the development of eDemocratic environments that allow better management and more citizen engagement.

The proposal is based on the use of "Tokens" to increase and reward the participation of the citizen and to promote the co-management of public utility projects.

It is aimed at finding new forms of commitment that can strengthen local democracies, as the technological deployment in the past decade has not been accompanied by greater participation or by a greater resilience of our democracies to the spaces of uncertainty generated by our society.

In Sect. 2, different solutions have been presented that corroborate the technological possibility of carrying out new steps in this direction, in order to address the existing deficit of citizen governance.

The blockchain technology applied to governance processes is yet to be developed. Therefore, there is an opportunity to experiment with new tools and new eParticipation formats to help consolidate and reform our democratic systems and to make them stronger and more participatory.

Currently, we are working on the development of a participation system based on this proposal for the City of Granada. We have the participation of the MediaLab of the University of Granada which is part of the project of the Integrated Sustainable Urban Development that must develop a participatory project as an urban innovative action.

We plan to use the UTAUT model (Unified Theory of Acceptance and Use of Technology) [28] to understand how this new eParticipation model will be valued among the citizens who will use it.

In parallel, we are the coordinators of the Working Group of the Association of Spanish Scientific and Technological Parks (APTE) that will launch the first permissioned blockchain network in the service of public research centers and R & D companies that are established in its network (53 parks), in order to develop new management processes, new dApps and disruptive solutions in different markets.

In both projects the "participatory tokens" are a strategic element for the results that are expected. These results will be the indicators of the degree of success of the model presented in this work, in the 2020 horizon.

References

1. Nakamoto, S.: Bitcoin: a peer-to-peer electronic cash system (2008). https://bitcoin.org/files/bitcoin-paper/bitcoin_es.pdf. Accessed 7 Feb 2019
2. Hedera hashgraph plattform. https://www.hedera.com/platform. Accessed 7 Feb 2019
3. IOTA foundation tangle protocol. https://www.iota.org/research/meet-the-tangle. Accessed 7 Feb 2019
4. Alexopoulos, C., Charalabidis, Y., Androutsopoulou, A.: Benefits and obstacles of blockchain applications in E-Government. In: Proceedings of the 52nd Hawaii International Conference on System Sciences, pp. 3377–3386 (2019)
5. Huang, J., Karduck, A.: A methodology for digital government transformation. J. Econ., Bus. Manag. 5(6), 246–254 (2017)
6. Morabito, V.: Blockchain governance. In: Business Innovation Through Blockchain: The B³ Perspective, pp. 41–59. Springer (2017)
7. Glaser, F., et al.: Blockchain as a platform. In: Treiblmaier, H., Beck, R. (eds.) Business Transformation through Blockchain, vol. I, Pp. 121–143. Palgrave MacMillan (2019)

8. Springall, D., et al.: Security analysis of the estonian Internet voting system. https://estoniaevoting.org/downloads/. Accessed 25 Mar 2019
9. Rauchs, M., et al.: Distributed ledger technology systems: a conceptual framework. Cambridge Centre for Alternative Finance (2018)
10. IIIT Hyderabad: Cloud computing for e-Governance, White Paper (2010). http://search.iiit.ac.in/uploads/CloudComputingForEGovernance.pdf. Accessed 8 Feb 2019
11. Bitnation Governance 2.0 Platform. https://tse.bitnation.co/. Accessed 9 Feb 2019
12. Follow My Vote Platform. https://followmyvote.com/. Accessed 9 Feb 2019
13. BoardRoom Platform. http://boardroom.to/#About. Accessed 9 Feb 2019
14. CONSUL Project. http://consulproject.org/en/index.html. Accessed 9 Feb 2019
15. TIVI Online Voting System. https://tivi.io/. Accessed 9 Feb 2019
16. Fernández, M.: Descifrar las smart cities: Qué queremos decir cuando hablamos de smart cities, pp. 183–185, Caligrama Editorial, Madrid (2016)
17. Romero-Frías, E., Arroyo-Machado, W.: Policy labs in Europe: political innovation, structure and content analysis on Twitter. In: El profesional de la información, 27(6), 1181–1192 (2018)
18. LabIn Granada: Social Innovation Lab. https://labingranada.org/. Accessed 10 Feb 2019
19. Garriga-Portolá, M., López Ventura, J.: The role of open government in smart cities. In: Open Government Public Administration and Information Technology 4. Springer, New York (2014)
20. Zhao, Z., Chan, T.H.H.: How to vote privately using bitcoin. In: International Conference on Information and Communications Security, pp. 82–96. Springer (2015)
21. Hopwood, D., et al.: Zcash protocol specification. Technical Report 2016-1.10, Zerocoin Electric Coin Company (2016)
22. McCorry, P., Shahandashti, S.F., Hao, F.: A smart contract for boardroom voting with maximum voter privacy. IACR Cryptology ePrint Archive 2017, 110 (2017)
23. Yu, B., et al.: Platform-independent secure blockchain-based voting system. In: Chen, L., Manulis, M., Schneider, S. (eds.) Information Security, ISC. Lecture Notes in Computer Science, vol. 11060, pp. 369–386. Springer, Cham (2018)
24. Annae, R., Freeland, R., Theodoropoulos, G.: E-voting requirements and implementation. In: 2007 The 9th IEEE CEC/EEE 2007, pp. 382–392. IEEE (2007)
25. Wang, K.H., et al.: A review of contemporary e-voting: requirements, technology, systems. In: Data Science and Pattern Recognition, vol. 1, no. 1, pp. 31–47 (2017)
26. Ayed, A.B.: A conceptual secure blockchain-based electronic voting system. Int. J. Netw. Secur. Appl. 9(3) 1–9 (2017)
27. Hardwick, F.S., Gioulis, A., Akram, R.N.: E-voting with blockchain: an e-voting protocol with decentralisation and voter privacy. ISG-SCC, Royal Holloway, University of London, UK (2018)
28. Naranjo-Zolotov, M., et al.: Examining social capital and individual motivators to explain the adoption of online participation. Future Gener. Comput. Syst. 92, 302–311 (2018)

Blockchain Approach to Solve Collective Decision Making Problems for Swarm Robotics

Trung T. Nguyen$^{(\boxtimes)}$, Amartya Hatua$^{(\boxtimes)}$, and Andrew H. Sung$^{(\boxtimes)}$

School of Computing Sciences and Computer Engineering,
The University of Southern Mississippi, Hattiesburg, USA
{trung.nguyen,amartya.hatua,andrew.sung}@usm.edu

Abstract. Recently, there are significantly increasing number of applications of swarm robotics for example targeted material delivery, precision farming, surveillance, defense and many other areas. Some qualities such as robot autonomy, decentralized control, collective decision-making ability, high fault tolerance etc. make swarm robotics suitable for solving these real-world problems. Blockchain, a decentralized ledger which is managed by a peer-to-peer network with cryptographic algorithms, provides a platform to perform different transactions in a secure way. The decentralized nature of swarm robotics makes it compatible to combine with blockchain technology and allows it to implement different decentralized decision making, behavior differentiation and other business models. This paper proposed a new distributed collective decision-making algorithm (best-of-n) for swarm robotics, where robots form a peer-to-peer network and perform different transactions using blockchain technology. For performance comparison, both collective decision-making algorithm with and without using blockchain have been implemented and their results have also been compared with respect to different metrics such as consensus time and exit probability. The performance shows that our proposed method gives a lower consensus time and higher exit probability value and therefore outperforms classical methods.

Keywords: Swarm robotics · Blockchain · Collective decision · Best-of-n

1 Introduction

Collective decision making [3–6], is defined as individuals in a group collectively make a decision from a number of alternatives without centralized leadership. Collective decision making [1,2] has normally been applied in two classes of problem: consensus achievement and task allocation. Comprehension of different collective decision making systems from nature and their design and implementations in the artificial system lead to the development of swarm robotics system [7,8]. This article will focus on solving collective decision making (best-of-n) [2] problems in swarm robotics system. Swarm robotics can be used in many

© Springer Nature Switzerland AG 2020
J. Prieto et al. (Eds.): BLOCKCHAIN 2019, AISC 1010, pp. 118–125, 2020.
https://doi.org/10.1007/978-3-030-23813-1_15

applications such as: autonomous army, disaster rescue missions, autonomous surveillance, plume tracking, cooperative environment monitoring, moving target localization and tracking. Based on quality and associated cost, best-of-n problems are mainly classified in five different sub-problems, discussed in Sect. 3.1. A decentralized distributed swarm robotics decision system has a few limitations. Firstly, the classical decision-making approaches use a predefined threshold time to check for consensus [10]. So sometimes the predefined time may be either insufficient or inefficient. Secondly, there is always a chance that a false positive result can occur, in which the swarm robotics system choose a non-optimal option [10].

Blockchain [16], a decentralized ledger which is managed by a peer-to-peer network with cryptographic algorithms, provides a platform to perform transactions in a secure way. Blockchain is a chain of blocks which contain data, hash of the block and hash of previous block. Using hash is not enough to secure the blockchain. Therefore, proof-of-work [17] is introduced for blockchain. This mechanism makes blockchain very secure, because if one block is tempered with, then all the following blocks' proof-of-works need to be calculated. All the nodes in the network create a consensus in which they agreed upon which blocks are valid and which are not. This mechanism helps to secure the data on blockchain. Another important part of the blockchain is a smart contract. A smart contract [15] is a computer code written mostly in Solidity. Every node of the blockchain network executes the smart contract to perform certain transactions and updates the distributed ledger.

The main objective of this research is to implement both the classical approach and blockchain approach on five scenarios of the best-of-n problems, and comparing their performance with respect to the consensus time and exit probability. In our experiment, we used a smart contract [15] to collect and count the votes cast by different agents or swarm robots.

2   Methodology

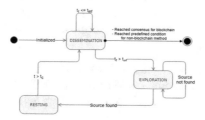

Fig. 1. State diagram of proposed methodology

The best-of-n problem for swarm robotics will be simulated with the following strategies: (1) The robots will move freely in a simulated environment which is bounded by four walls. (2) The environment contains two areas:

- REST_AREA: where the robots stay at the beginning or return after finding the light sources (denote the possible options in the best-of-n problem).
- EXPLORATION_AREA: where the robots can find the light sources (LIGHT_SRC). The option is denoted by the light sources, and each light source will have different quality (intensity of the light source) and different cost (the distance from REST_AREA to the light source).

INITIAL STATE: In this state, the robots are scattered in REST_AREA. Moreover, stopping conditions need to be set in this state. For non-blockchain implementation, stopping condition is based on the predefined maximum running time and predefined maximum number of unchanged opinions. For blockchain implementation, stopping condition is determined by whether consensus state is reached by all robots.

DISSEMINATION STATE: If the stopping conditions are not met, then the agents will use the influencing factor time (t_{inf}), which is calculated based on the quality and the exploration cost to obtain the last opinion, to communicate with other agents about their opinions. During this time, the agents affect neighboring agents and collect the opinion of the neighboring agents and they perform Direct Modulation of Majority-based Decisions (DMMD) [14]. For the blockchain approach, every agent casts a vote as their opinion and all the votes are managed by the smart contract in the blockchain. Once the agents choose an option among n options, they change their state to EXPLORATION state and continue to explore that option until the stopping conditions are satisfied.

EXPLORATION STATE: In this state, the agents will move from REST_AREA to EXPLORATION_AREA for finding the light source using their sensor. Once the agents find the source (i) and measure the quality (q_i), they start moving back the REST_AREA. The exploration time is calculated by the summary of the time to find the source (c_{ij}^1) and the time to return to the REST_AREA (c_{ij}^2). The influence time factor will then can be calculated based on the quality and the exploration cost of option i using the following equation: $w_1 * q_i + w_2 * (C - c_{ij}^1 - c_{ij}^2)$ [where C is constant value to which is introduced to calculate the effect of exploration cost]. After returning to the REST_AREA, the agents stay in NEST state for a constant amount of time (t_c) then move to the DISSEMINATION state.

NEST STATE: The robot agents will be in NEST state after they explore the environment, find the light source, and return to REST_AREA. During this state, the agents will collect other agents' opinions to decide later in DISSEMINATION state.

FINISH STATE: The state when the stopping conditions are reached, or the robot agents reach to consensus state through blockchain smart contracts.

3   Implementation

3.1   Experimental Setup

To perform the experiments for our proposed algorithm and classical method, ARGOS 3 swarm robotics simulator [11] and the ARGOS-Footbot [8] plugin. In our experiments, two options ($n = 2$) were only used. Below is the experiment setup information for five scenarios of best-of-n problems:

Same Quality Same Cost (SQSC): This type of experiment scenario is followed the experimental setup described by Brambilla et al. [12].

Same Quality Different Cost (SQDC): This type of experiment scenario is followed the experimental setup described by Scheidler et al. [13].

Different Quality Same Cost (DQSC): In this experiment scenario, the black colored region is the resting area and the two white regions are the exploration area. The light sources are different in quality in either of the exploration areas but their distance from the REST_AREA is equal.

Different Quality Different Cost (Synergic): The map is used for this experiment has one rest area and two exploration areas, but the distance between the rest area and the high-quality source area is less than that of the rest area and the low-quality source area.

Different Quality Different Cost (Antagonistic): This type of experiment scenario is similar to the previous one. However, the rest area is nearer to low-quality source area and farther from the higher quality source and exhibits the antagonistic property. In this case, the agents need to select an optimized decision, which is the trade-off between quality and cost.

3.2   Non-blockchain Implementation

With the non-blockchain approach, the majority voting (DMMD) decision making strategy is used by the robot agents to decide whether keeping the current opinion or changing to the majority opinion among its neighbors. There are two stopping conditions: the maximum running time and the minimum unchanged opinion times. If one robot continuously doesn't change its opinions for a number of times, that robot can assume that consensus is reached and switch to STOPPING state. The experiment is considered to reach to FINISH state when all of the robots in STOPPING state. Figure 2 displays the detailed flowchart to implement Non-blockchain approach.

3.3   Blockchain Implementation

Each robot keeps a separate copy of the blockchain and acts as a full node and miner in the blockchain network. The following step will describe how blockchain is used to setup and employ for our best-of-n experiments.

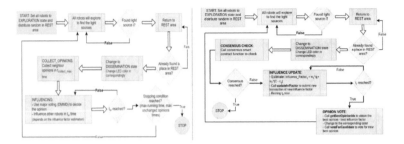

Fig. 2. Block diagram of Non-blockchain (left) and Blockchain (right) implementation

Initialization: We employ Go Ethereum implementation library (Geth) to setup the private Ethereum network and deploy smart contracts. To initialize Geth nodes on private network, a custom genesis file needs to be defined. In this custom genesis file, a fixed small mining difficult (1000) is used which is suitable for the limited computing power of robots. At first, a miner node will be initialized, submit our smart contracts, perform mining jobs to successfully deploy the smart contracts to the private Ethereum network, and finally obtain the contract address. Nevertheless, each robot will start its own Geth node (process) and connect to the private Ethereum network by calling **add_peer** method with its **enode** information. After the smart contract is successfully deployed and the contract address is obtained, the miner node will send the contract address to all robots Ethereum nodes.

Smart Contracts: Smart contract is a programmed code that will be executed and verified by every node of the blockchain network.

Robot Controlling Routines: For the non-blockchain approach, we keep most of the non-blockchain implementation but make some changes. Firstly, after returning to the REST area, each robot will change to Dissemination state D_i in corresponding with the new opinion i. Each robot will call the smart contracts function named **consensus** (proof-of-work) to check whether or not the consensus state is reached by all robots. If it is, then the robot will switch to STOPPING state. If consensus state is not reached, robot j can estimate the influence factor of the new opinion i of robot j by the equation below:

$$Influence\_Factor_{ij} = w_1 * q_i + w_2 * (C * c_{ij}) \qquad (1)$$

where $w_1; w_2$ denote the positive weight for quality and cost correspondingly ($0 \leq w_1; w_2 \leq 1$), i is the quality value of opinion i; c_{ij} denotes the exploration cost for robot j to find source i; C is the constant to denote the max exploration time. Secondly, unlike non-blockchain approach, each robot will have to use smart contracts function named **updateInfFactor** to submit new transaction of its new influence factor so that everyone can compare and select the best one. Here, we keep a fixed amount of time (t_r) for every robot who has new opinion to submit their new information and wait for new information is sharing around (through blockchain synchronization). After the (t_r) is over, each robot will call smart contracts function named **getBestOpinionIdx** to obtain the best opinion which has the best influence factor at that time. That robot will change its color in corresponding to the best opinion. Then it will need to call smart contracts function named **voteForCandidate** to vote for the new best opinion. Finally, it will switch to Exploration state to continue exploring to update influence factor in next round.

3.4 Performance Metrics

To measure the performance of the non-blockchain and blockchain approached, two following metrics are used:

Consensus Time (T_consensus): The number of seconds until all robots reached to same opinion.

Exit Probability (E_prob): Exit probability is defined as "the probability to make the best decision, computed as the proportion of runs that converge to consensus on opinion a" [14]. In our experiments, "a" is defined as best opinion which reflects the trade off between quality and cost.

4 Result and Discussion

Every experiment is performed twenty-five times. The consensus time and exit probability results are mentioned in the following subsections.

4.1 Consensus Time

In every case, the non-blockchain experiment has taken more time than corresponding blockchain experiment to reach the consensus state. As the complexity of the experiment increased, especially for different quality different cost (synergic and antagonistic) experiment, the difference in consensus time between blockchain method and non-blockchain method are very high. A statistical description of consensus time data is presented in Table 1. In Table 1, it is clearly visible that in all the cases with respect to all the parameters, the experiment with blockchain implementation outperforms its non-blockchain counterpart.

4.2 Exit Probability

According to the collected results in Table 2, the non-blockchain swarm robotics chooses the non-optimal alternative in many occasions, while the experiments with blockchain don't have any false positive case. This is another advantage of blockchain in swarm robotics system besides low consensus time for a collective decision making problem.

Table 1. Statistical description of consensus time data

Type of Best-of-N problem	Implementation type	Minimum	Q1	Median	Q3	Maximum	Mean	Range
Same quality same cost	Non blockchain	5.07	54.21	79.74	107.02	393.59	96.43	388.51
	Blockchain	29.95	38.55	45.89	53.94	69.70	46.51	39.76
Same quality different cost	Non blockchain	73.61	106.56	143.77	197.1	1376.1	188.66	1302.56
	Blockchain	50.61	83.56	120.73	174.13	1353.18	165.66	1302.56
Different quality same cost	Non blockchain	59.60	108.76	134.21	161.55	448.12	150.96	388.15
	Blockchain	47.37	56.00	63.34	71.39	87.15	63.96	39.77
Synergic	Non blockchain	254.96	334.19	399.20	526.05	1165.6	452.44	910.64
	Blockchain	43.84	71.28	138.89	182.95	335.54	137.29	291.70
Antagonistic	Non blockchain	122.85	272.24	459.40	626.32	1946.81	521.27	1823.95
	Blockchain	37.95	194.48	230.83	279.69	562.16	232.14	524.21

Table 2. False positive and exit probability of the experiments

Exit probability	Same quality same cost	Same quality different cost	Different quality same cost	Synergic	Antagonistic
Non blockchain	0.88	0.8	0.88	0.96	0.72
Blockchain	1	1	1	1	1

5   Conclusion

In this paper, a new collective decision making algorithm using blockchain technology has been proposed for solving best-of-n problems of swarm robotics. To evaluate the effectiveness of our proposed algorithm, we have implemented the new algorithm (using blockchain) and classical algorithms (without using blockchain). We then conducted experiments to perform two algorithms on different variants of best-of-n problems. The results are compared on two types of metrics: consensus time and exit probability. Compared to non-blockchain methods, our recommended algorithm requires less time to reach consensus status. Secondly, while classical methods produce false positive results, our proposed algorithms have chosen the optimal alternative in every situation. Moreover, blockchain can provide a more secure communication platform between swarm robotics agents. In conclusion, our proposed collective decision making algorithm using blockchain technology can help us to achieve the most optimal results in less time (lower consensus time), higher exit probability (no false positive) and more secure way. In future, this work can be extended by implementing the new algorithm on real robot platforms and perform experiments on different real environment setups. Other than that, other collective decision making problems such as obstacle avoidance, collective map generation, task allocation, etc. can also be solved using blockchain technology.

References

1. Bose, T., Reina, A., Marshall, J.A.: Collective decision-making. Curr. Opin. Behav. Sci. **16**, 30–34 (2017)
2. Valentini, G., Ferrante, E., Dorigo, M.: The best-of-n problem in robot swarms: formalization, state of the art, and novel perspectives. Front. Robot. AI **4**, 9 (2017)
3. Okubo, A.: Dynamical aspects of animal grouping: swarms, schools, ocks, and herds. Adv. Biophys. **22**, 1–94 (1986)
4. Sumpter, D.J.: Collective Animal Behavior. Princeton University Press, Princeton (2010)
5. Kao, A.B., Miller, N., Torney, C., Hartnett, A., Couzin, I.D.: Collective learning and optimal consensus decisions in social animal groups. PLoS Comput. Biol. **10**(8), e1003762 (2014)
6. Strandburg-Peshkin, A., Farine, D.R., Couzin, I.D., Crofoot, M.C.: Shared decision-making drives collective movement in wild baboons. Science **348**(6241), 1358–1361 (2015)
7. Brambilla, M., Ferrante, E., Birattari, M., Dorigo, M.: Swarm robotics: a review from the swarm engineering perspective. Swarm Intell. **7**(1), 1–41 (2013)
8. Sonawane, P., Nyalpelly, O., Talele, M., Modi, R., Lade, N.: Android based master slave robot. Environment **14**, 2
9. Bernstein, D.S., Givan, R., Immerman, N., Zilberstein, S.: The complexity of decentralized control of Markov decision processes. Math. Oper. Res. **27**(4), 819–840 (2002)
10. Ferrer, E.C.: The blockchain: a new framework for robotic swarm systems (2016). arXiv preprint arXiv:1608.00695
11. Valentini, G., Brambilla, D., Hamann, H., Dorigo, M.: Collective perception of environmental features in a robot swarm. In: International Conference on Swarm Intelligence, pp. 65–76. Springer, Cham, September 2016
12. Brambilla, M., Brutschy, A., Dorigo, M., Birattari, M.: Property-driven design for robot swarms: a design method based on prescriptive modeling and model checking. ACM Trans. Auton. Adapt. Syst. **9**(4), 17 (2015)
13. Scheidler, A., Brutschy, A., Ferrante, E., Dorigo, M.: The k-unanimity rule for self-organized decision-making in swarms of robots. IEEE Trans. Cybern. **46**(5), 1175–1188 (2016)
14. Dorigo, M., Birattari, M., Brambilla, M.: Swarm robotics. Scholarpedia **9**(1), 1463 (2014)
15. Buterin, V.: A next-generation smart contract and decentralized application platform. White paper (2014)
16. Swan, M.: Blockchain: Blueprint for a New Economy. O'Reilly Media Inc., Sebastopol (2015)
17. Vukoli, M.: The quest for scalable blockchain fabric: proof-of-work vs. BFT replication. In: International Workshop on Open Problems in Network Security, pp. 112–125. Springer, Cham (2015)

Prediction of Transaction Confirmation Time in Ethereum Blockchain Using Machine Learning

Harsh Jot Singh[✉] and Abdelhakim Senhaji Hafid

Montreal Blockchain Lab, University of Montreal, Montreal, QC, Canada
harsh.jot.singh@umontreal.ca, ahafid@iro.umontreal.ca

Abstract. An Ethereum transaction is defined as the method by which the external world interacts with Ethereum. More and more users are getting involved in cryptocurrencies like Ethereum and Bitcoin. With a sudden increase in the number of transactions happening every second and the capital involved in those transactions, there is a need for the users to able to predict whether a transaction would be confirmed and if yes, then how much time would it take for it to be confirmed. This paper aims to use modern machine learning techniques to propose a model that would be able to predict the time frame within which a miner node will accept and include a transaction to a block. The paper also explores the impact of imbalanced data on our chosen classifiers-Bayes, Random Forest and Multi-Layer Perceptron (MLP) with SoftMax output and the alternative performance measures to optimally handle the imbalanced nature of the dataset.

Keywords: Ethereum · Transactions · Machine learning

1 Introduction

Since the emergence of Bitcoin, which at grassroots, was meant to replace the state-backed currencies with a digital version that not only integrates cryptography, networking and opensource technologies but is also meant to cross international boundaries and nullify the need of banks as a mean to store money [1], there has been a gradual increase in the awareness and excitement in regards to cryptocurrencies and the rudimentary distributed ledger (or Blockchain) technology. The market capitalization of publicly traded cryptocurrencies is currently about $122.1 billion. These currencies offer users pseudo-anonymity with respect to their personal information and at the same time allows the transactions data to be publicly available for analysis and/or characterization of its users [2]. Ethereum, the second most important cryptocurrency, after Bitcoin, provides formidable functionalities in comparison to its counterparts. These added functionalities allow for the Ethereum dataset to be utilized for analysis and knowledge inference. As such, in this paper, we sought out to gather the comprehensive Ethereum transaction data and the associated metadata to predict the time frame within which a mining node will confirm a transaction based on its associated metadata.

© Springer Nature Switzerland AG 2020
J. Prieto et al. (Eds.): BLOCKCHAIN 2019, AISC 1010, pp. 126–133, 2020.
https://doi.org/10.1007/978-3-030-23813-1_16

1.1   Ethereum

Ethereum is a project developed to facilitate transactions among individuals without them needing to divulge personal details. It helps develop a trusted relationship between two anonymous users. It is a transaction-based state machine that starts from the genesis state [3] and reaches the final state by incrementally executing a transaction [4]. There can be valid or invalid intermediate changes. Invalid state changes could imply an invalid account balance modification in either of the two accounts; sender or receiver. Formally [5],

$$\sigma_{t+1} \equiv Y(\sigma_t, T)$$

where, Y is Ethereum state transition function, which allows components to carry out computations and σ allows them to store arbitrary state between transactions.

Merkel trees [6] are used to combine multiple transactions in a block, which work as journals recording the transaction. The blocks are connected one another using a cryptographic hash.

The basic unit of Ethereum is account. Each account has a 20-byte address. For an individual to make a transaction they need to have an account. These Ethereum accounts are of two types: Externally Owned Account (EOA) or Contract Accounts. EOA are controlled by private keys whereas Contract Accounts are controlled by their contract code and can only be activated by an EOA.

1.2   Machine Learning

Machine Learning provides explicit learning capabilities to a computing system by using statistical techniques. A ML algorithm takes a set of input samples called training set and uses the set to learn in three fundamental ways: supervised, un-supervised and reinforcement learning. In supervised learning, the algorithm also has knowledge of the target output for the training set, known as labels, and the algorithm learns the input to achieve the target. There are no such labels in un-supervised learning. The reinforcement learning deals with the problem of learning the appropriate action(s) in order to maximize payoff. This paper only employs supervised learning techniques.

Due the volatile nature of the Ethereum transaction dataset, its predictability has been sparsely covered in published literature. However, Bhat and Vijayal [7] provided a brief analysis of blockchain technologies and technical aspects of Ethereum and Bitcoin while also discussing the mining approach used to validate transactions and to create new coins. The comparison between Ethereum and Bitcoin concluded that Ethereum is far more extensible and secure that its counterpart. Chen, Narwal and Schultz [8] proposed a model for prediction on price changes in Ethereum. They employed price data sampled at one-hour intervals and utilized machine learning models such as Support Vector Machines (SVM), Naïve Bayes, Random Forest etc. to predict the prices resulting in the maximum prediction accuracy of 61.12%.

The remainder of the paper is organized as follows. Section 2 presents the acquired dataset and the three machine learning models used for prediction purposes. Section 3 presents the performance evaluation criteria. Section 4 evaluates the performance of the three models. Section 5 concludes the paper and presents future work.

2   Proposed Model

In this section, we present the proposed methodology for the prediction problem. We first discuss the dataset and the associated features, its imbalanced nature and how it was utilized to solve the problem. We then briefly discuss the machine learning models used in order to predict the confirmation time for the transactions.

2.1   Data Set and Features

The primary dataset consists of a static block of a million transactions until November 18th, 2018. We extracted the dataset from the Ethereum.io API [9]; it includes the number of transactions made by the sender, the gas provided by the sender, gas price, gas used, the timestamp for when the transaction was sent and the block timestamp to which it belongs. While there is no explicit field in the Ethereum API to calculate timestamp of when a transaction was sent, we wrote a script to access the pending transaction pool [10] and to record the timestamp of when a transaction was added to the pool ('Last Seen'). The block timestamp represents the time when the transaction was confirmed by a mining node. The difference between these two timestamps represents the time taken to confirm the transaction.

We classified the confirmation time in 8 categories: within 15 s (or approximately 1 block time), within 30 s, within 1 min, within 2 min, within 5 min, within 10 min, within 15 min, within 30 min or longer. Most of the transactions (49.4%) belonged to class one and about 1% (see Fig. 1) of these transactions were reverted transactions or did fail. This can be visualized in Table 1.

Table 1. Statistics of dataset

Class	Number of transactions
1. Within 15 s	494,071
2. Within 30 s	248,609
3. Within 1 min	173,047
4. Within 2 min	38,918
5. Within 5 min	33,951
6. Within 10 min	31,349
7. Within 15 min	8,934
8. Within 30 min and longer	1,121

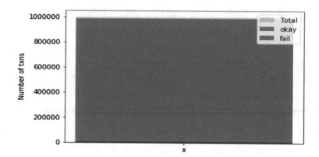

Fig. 1. Failed to total transactions

SMOTE Algorithm. The algorithm was used to handle the imbalanced nature of the dataset. It is an oversampling technique which creates synthetic examples of the minority class in an imbalanced dataset. These examples are variations of real data with some operation done on them to distinguish them from the real samples instead of creating multiple copies of real minority data like in oversampling techniques. The sklearn imbalanced-learn toolbox was used to apply SMOTE [11].

Training Plan. Once data was obtained, each dataset was split into 2 disjoint sets. The first set which comprises 80% of the data was used for training and validation. The second set was used as a test data set in order to compare the performance of each of the three models on identical data instead of a randomly split dataset.

2.2   Methods

We now discuss the three learning models employed in this paper. The dataset (as discussed in Sect. 2.1) was executed on multiple supervised learning models such as kNN classifier, Support Vector Machine (SVM), Naïve Bayes, Random Forest etc. But the results with kNN and SVM were too inaccurate to include in the paper.

Naïve Bayes. Given a new, unseen data point $z = (z_1, \ldots, z_M)$, naive Bayes classifiers, which are a family of probabilistic classifiers, classify z based on applying Bayes' theorem with the "naive" assumption of independence between the features (attributes) of z given the class variable t. By applying the Bayes' theorem, we have [12]:

$$p(t = c | z_1, \ldots, z_M) = \frac{p(z_1, \ldots, z_M | t = c) p(t = c)}{p(z_1, \ldots, z_M)}$$

which can be simplified as:

$$y = \arg \max p(t = c) \prod_{j=1}^{M} p(z_j | t = c)$$

Random Forests. Arandom forest is a ML technique which operates on the principle of creating multiple decision trees which are all trained with variations of the training dataset. The output of model is the mode of the output (classification) or mean prediction (regression) of each of these 'n' decision trees.

Instead of decision trees, new models comprising of multinomial logistic regression and naïve bayesian classifiers have also been introduced to achieve better predictive performance. This paper uses a random forest classification model.

Multi-layer Perceptron. A MLP is a feed-forward neural network with at least one hidden layer. Each layer, except the input layer, of MLP has a non-linear activation function. The model employs supervised learning with back-propagation for the training phase. The output layer of each model has a non-linear function that classifies an example into a class in a given set of classes.

An MLP with batch normalization [13] and dropout [14] is used for the dataset with a SoftMax output layer, and Rectified linear units as the non-linearities between layers. Cross entropy was used as the loss function for the network. Optimization of the neural network was performed using AdaGrad optimization [15]. Furthermore, the learning rate was decayed exponentially with each epoch of training.

3   Evaluation Criteria

It is important to evaluate the strength of any given model especially if the dataset involved is imbalanced and could lead to inaccurate results. This section gives a brief description about the evaluation criteria employed to measure the performance of the three models.

3.1   Accuracy

With imbalanced data, one must face the accuracy paradox. If a class is abundant in a dataset, the model may be tempted to always classify each entry with this class. The accuracy may be high, but the model may not be able to find the entries of the under-represented class(es). In those cases, the accuracy must be compared with the null-accuracy, which is a metric defined as the accuracy of our model if it always predicted the most represented class for each entry.

Table 2. Interpretation of Cohen's Kappa score.

Value of Kappa	Level of agreement	% of data that is reliable
0–0.20	None	0–4%
0.21–0.39	Minimal	4–15%
0.40–0.59	Weak	15–35%
0.60–0.79	Moderate	35–63%
0.80–0.90	Strong	64–81%
Above 0.90	Almost perfect	82–100%

3.2   Cohen's Kappa

Cohen's Kappa [16] tells us how much better our model is performing over the performance of another model which will simply guess the result based on the frequency of each class. A value of 0 indicate that the classifier is useless. The interpretation of the Cohen's Kappa score is given in Table 2.

4   Performance Evaluation

The three models were first evaluated with the static data (as discussed in Sect. 3.1) and then, they were also evaluated as incrementally re-trained models. This section presents the results observed with the three models when used on both static and incrementally updated real time dataset.

4.1   Initial Performance

During the evaluation process of the three models, a randomly selected data subset of 100,000 transactions was used and the prediction accuracy, null accuracy and Cohen's kappa score for each model was calculated.

It should be noted that while the data acquired from the Ethereum API had several attributes, they did not all have the same effect on the prediction accuracy. It was observed that the gas price, gas limit and nonce have the most effect on the confirmation time for a transaction and that the model accuracy can be increased by a large factor if weighted models are employed. Out of three, only MLP offers weighted nodes for parameters. The other two model were modified as locally weighed Naïve Bayes [17] and weighted random forest [18]. The evaluation results can be observed in Table 3.

Table 3. Initial evaluation results

Training data	Time to train (in minutes)	Accuracy	Null accuracy	Cohen's Kappa	Model used
100,000	15	99.54%	98.85%	26.74%	Naïve Bayes
100,000	20	97.39%	97.46%	85.00%	Random Forest
100,000	53	85.27%	55.32%	73.61%	MLP
1,000,000	25	92.08%	90.5%	27.70%	Naïve Bayes
10,00,000	40	96.78%	95.13%	83.42%	Random Forest
10,00,000	130	92.91%	67.39%	78.50%	MLP

The naïve Bayes model resulted in best prediction accuracy however, the null accuracy for the model was also very similar to the prediction accuracy hence rendering the model inaccurate. The random forest model came out with the best Cohen's kappa score but also had high null accuracy relative to prediction accuracy hence making MLP the best but moderately accurate model for prediction of the classes for the testing data.

The three models were then executed with the whole dataset providing similar results but with better accuracy and Cohen's Kappa score implying that these models are better trained with more data available to train.

4.2   Real-Time Dataset

In order to accurately predict the confirmation time for a transaction, the model needs to keep the network congestion in mind. Hence there is a need for the model to be re-trained incrementally. In this paper, this was only implemented to the MLP model.

To train the model from scratch at each interval would prove to be inefficient due to high training time. To make it work, after training the network, the weights associated with it are saved to the disk and loaded when new data becomes available. This allows the model to be trained from where it was last trained.

According to the data collected from Ethstats.net [19], each block in Ethereum has a gas limit of 7,999,992 and each transaction costs 21,000 gas, assuming nothing else is attached to it and that the mining nodes are in fact generating full blocks. This implies that there are ~ 380 transactions (upper limit) in each block with a block time of ~ 15.03 s. This would result in about 25.346 txn/s.

The MLP model proposed in Sect. 3.2 was re-trained at 10, 15 and 30 min intervals and following results were observed:

Table 4. Real-time data evaluation results.

Time interval (in minutes)	No. of transactions (thousands)	Time to train (in minutes)	Model accuracy	Cohen's Kappa
10	15.6	~ 14	83.61%	78%
15	23.4	~ 15	82.34%	78.12%
30	46.8	~ 17	82.18%	78.77%

5   Conclusion and Future Work

Data collection and handling the imbalanced nature of the real-time data is challenge and the task of developing a model with low training time and better prediction accuracy is non-trivial. This paper proposed three different models to predict the execution time of an Ethereum transaction by classifying them into five classes based on historical data. It was observed that while a model might have great accuracy, it might not be a strong model [16] due to the imbalanced nature of the data involved. The study presents results that indicate that of all the three models proposed, the MLP with SoftMax output function is the only one that provided moderately accurate results given the volatility and imbalanced nature of the dataset.

Looking forward, it would be prudent to develop a model that could do these predictions with lower execution time and better accuracy. The idea of predicting block confirmation time for an ethereum transaction is relatively new and should be explored further to utilize regression methods for prediction. While classification, as observed, has provided quite good results, it would be interesting to find out which of the two would be more accurate for block confirmation time predictions. There is also the possibility of the model to be able to provide insight to what changes to the transactions a user could make to make sure it is confirmed within a block time.

References

1. Greenberg, A.: https://www.forbes.com/forbes/2011/0509/technology-psilocybin-bitcoins-gavin-andresen-crypto-currency.html#7c44500c353e. Accessed 31 Jan 2019
2. Payette, J., Schwager, S., Murphy, J.: Characterizing the Ethereum Address Space. http://cs229.stanford.edu/proj2017/final-reports/5244232.pdf. Accessed 02 Feb 2019
3. Genesis Block. https://en.bitcoin.it/wiki/Genesis_block. Accessed 02 Feb 2019
4. Saraf, C., Sabadra, S.: Blockchain platforms: a compendium. In: IEEE International Conference on Innovative Research and Development (ICIRD) 2018, Bangkok pp. 1–6 (2018)
5. Wood, G.: Ethereum: a secure decentralised generalised transaction ledger. https://gavwood.com/paper.pdf. Accessed 02 Feb 2019
6. Merkle Tree. https://en.wikipedia.org/wiki/Merkle_tree. Accessed 02 Feb 2019
7. Bhat, M., Vijayal, S.: A probabilistic analysis on crypto-currencies based on blockchain. In: International Conference on Next Generation Computing and Information Systems (ICNGCIS) 2017, Jammu, pp. 69–74 (2017)
8. Chen, M., Narwal, N., Schultz, M.: Predicting price changes in Ethereum. http://cs229.stanford.edu/proj2017/final-reports/5244039.pdf. Accessed 02 May 2019
9. https://etherscan.io/. Accessed 15 Jan 2019
10. https://etherscan.io/txsPending. Accessed 15 Jan 2019
11. https://imbalanced-learn.readthedocs.io/en/stable/api.html. Accessed 05 Feb 2019
12. Zhang, H.: The Optimality of Naïve Bayes. http://www.cs.unb.ca/~hzhang/publications/FLAIRS04ZhangH.pdf. Accessed 05 Feb 2019
13. Ioffe, S., Szegedy, C.: Batch normalization: accelerating deep network training by reducing internal covariate shift. Cornell University Library. arXiv:1502.03167, March 2015
14. Srivastava, N., Hinton, G., Krizhevsky, A., Sutskever, I., Salakhutdinov, R.: Dropout: a simple way to prevent neural networks from overfitting. J. Mach. Learn. Res. **15**(1), 1929–1958 (2014)
15. https://www.tensorflow.org/api_docs/python/tf/train/AdagradOptimizer. Accessed 05 Feb 2019
16. McHugh, M.: Interrater reliability: the kappa statistic. Biochem. Med. **22**(3), 276–282 (2012)
17. Frank, E., Hall, M., Pfahringer, B.: Locally Weighted Naïve Bayes. https://www.cs.waikato.ac.nz/~eibe/pubs/UAI_200.pdf. Accessed 05 Feb 2019
18. Winham, S., Freimuth, R., Biernacka, J.: A weighted random forest approach to improve predictive performance. Stat. Anal. Data Min. **6**(6), 496–505 (2013)
19. Ethereum Network Status. https://ethstats.net/. Accessed 15 Jan 2019

Improving Event Monitoring in IoT Network Using an Integrated Blockchain-Distributed Pattern Recognition Scheme

Anang Hudaya Muhamad Amin[1(✉)], Ja'far Alqatawna[1], Sujni Paul[1],
Fred N. Kiwanuka[1], and Imtiaz Ahmad Akhtar[2]

[1] Higher Colleges of Technology, Dubai, UAE
{aamin,jalqatawna,spaul,fkiwanuka}@hct.ac.ae
[2] IT Consultant, Stockholm, Sweden
cimtiaz@msn.com

Abstract. The application of blockchain technology for data storage and verification has been expanding from financial applications to other fields such as asset management and event monitoring in Internet-of-Things (IoT). This expansion consequently intensifies the problem of an increasing size of data stored in the blockchain, especially in event monitoring application where streams of data need to be stored and verified accordingly. In this paper, we propose an IoT-blockchain event monitoring framework that utilizes a distributed pattern recognition scheme for event data processing. Event data are treated as patterns comprising individual data retrieved from interconnected IoT sensors within a network composition. Preliminary results obtained indicate that the proposed scheme is capable of reducing the number of data blocks generated in the blockchain network, hence minimizing the needs for intensive storage and verification.

Keywords: Internet-of-Things (IoT) · Blockchain technology · Distributed pattern recognition · Event processing

1 Introduction

Industrial process monitoring and control is an important task that has been commonly carried out by network of Internet-of-Things (IoT) devices. Existing event monitoring schemes are still relying on a centralized processing, in which event data are collectively sent and analyzed at a processing node. This scenario may lead to the possibility of single-point failure for the entire monitoring network.

Pattern recognition for event monitoring has been adopted in different kinds of applications, including industrial automation and structural monitoring. With

© Springer Nature Switzerland AG 2020
J. Prieto et al. (Eds.): BLOCKCHAIN 2019, AISC 1010, pp. 134–144, 2020.
https://doi.org/10.1007/978-3-030-23813-1_17

the advancement in the field of IoT and distributed systems, the need for a distributed mechanism for pattern recognition is considered crucial [3]. Distributed pattern recognition algorithm with associative memory capabilities such as Hierarchical Graph Neuron (HGN) [11] and Vector Symbolic Architecture (VSA) [9] offers promising approach towards scalable event monitoring in a distributed processing environment.

In this paper, we focus on the data processing issue in IoT network, particularly in the application of event monitoring in a distributed environment. The existing centralized event monitoring schemes, such as in the works of Ahmad Jan et al. [7] suffer from several drawbacks when applied to the IoT-related applications. Firstly, centralized event data analysis may leads towards single-point of failure, in which event monitoring fails due to the failure of centralized server. Secondly, with massive number of IoT devices used, the amount of transmitted data to a centralized repository would be significantly increased. Thus, creating a connection bottleneck from IoT network to the central server.

To address the performance issue in regards to the number of data blocks generated in typical IoT-Blockchain network, we propose an event monitoring framework using an integrated blockchain-distributed pattern recognition scheme on IoT network. The main contributions of this paper are summarized as follows:

- Minimization of data block generation for event data storage and verification. By capturing and analyzing critical events involving entire network of IoT devices; this would limit the needs for block generation on blockchain network involving non-critical event.
- Scalable event monitoring scheme for large-scale network. With the use of associative memory-based distributed pattern recognition algorithm, detection of events from the bird's eye view perspective could be achieved by overlooking entire network for critical events.

The rest of the paper is organized as follows. Section 2 describes the IoT-blockchain integration from the perspective of distributed event monitoring and a review of existing IoT-blockchain implementation and other related concepts. Additionally, it discusses related concepts such as the distributed pattern recognition. In Sect. 3, we provide a description on the proposed design and implementation of a distributed pattern recognition scheme for event monitoring. Section 4 focuses on a simulation work carried out to simulate the proposed IoT-Blockchain event monitoring framework and preliminary results obtained. Finally, Sect. 5 concludes the paper and highlights some future directions.

2 IoT-Blockchain Integration and Related Components for Event Monitoring

The existence of blockchain technology can be traced back since the emergence of Bitcoin concept in 2008 [10]. Previous implementations of blockchain come with several limitations [12], including slow block generation period and lack of loop implementation (Turing incomplete) [4]. Figure 1 illustrate the decentralized blockchain approach in relation to the IoT environment, in comparison with existing centralized approach.

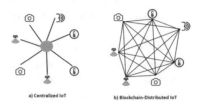

a) Centralized IoT   b) Blockchain-Distributed IoT

Fig. 1. IoT-blockchain network layout in comparison with existing default centralized IoT network.

Integrating blockchain into IoT network could be a challenge, as it requires fast validation and verification to be performed on the transactions within the network. Nevertheless, existing blockchain network such as Ethereum [6] has made this possible with its implementation of smart contract mechanism that allows transactions to be captured and verified.

2.1 "Mining" Operation in Blockchain

In blockchain implementation, a critical function that requires complex computation is the "mining" operation. This mining of hash value involves scanning for a value to be used with SHA-256 hashing algorithm to create a hash which has lesser value than a specific value. This value is used to identify different blocks being generated in a blockchain network [10].

In this paper, we propose a collective method of event data storage and verification on blockchain network using a distributed pattern recognition approach. Sensor data are treated as activation patterns as shown in Fig. 2. From the temporal perspective, this sensor data patterns could vary across time. Our interest would be to capture unique patterns to be stored and verified on blockchain network. Similar data patterns would be treated as non-event and will be discarded. Using this approach, our hypothesis is that the block generation could be minimized accordingly.

2.2 Hierarchical Graph Neuron (HGN) Learning Mechanism

The Hierarchical Graph Neuron (HGN) algorithm is an extended version of graph-based neural network scheme known as Graph Neuron (GN) [8]. GN-based algorithms basically implement a recognition based on the comparative neuron activation function. The firing value for each neuron depends upon the comparison function between input and stored elements within each particular neuron, which is known as bias entries. These bias entries are the *(value, position)* pair compositions that represent pattern elements. These entries were obtained from the synaptic responses between adjacent neurons (the firing values of adjacent neurons). The neuron firing mechanism in GN implementations is different from other neural-network techniques in the sense that its synaptic plasticity is independent of weight adjustment mechanism based upon the input strength. Rather, the value-comparison function between adjacent neurons is used to determine the output of each neuron.

Figure 3 shows the network composition of HGN for pattern input of length 5. HGN network in its actual formation, replicates the multiple layers of GN structure. The recognition procedure involves forward propagation approach without any backward propagation (from top neuron to the bottom-level neurons). The composition of neurons in each HGN network, N_{hgn} can be represented in the form of the following equation:

$$N_{hgn} = \left(\frac{p+1}{2}\right)^d \tag{1}$$

Where p and d represent the pattern size and dimension respectively.

Each neuron compares input pattern values with the information obtained from its adjacent neurons. Adjacency information for each neuron is stored in a data structure known as bias entry, which consists of the *(left, right)* value information. Each activated neuron therefore records the information retrieved from its adjacent left or right nodes. Collectively, these entries form a composition known as bias array. Each neuron's bias array only stores the unique adjacency information derived from the input patterns.

For a given N bias array size, the pattern storage process can be symbolically represented as follows:

$$E_B = \{\langle x, y\rangle; x \in N, y \in N\} \tag{2}$$

Where E_B, represents the sets of two-element ordered pair, while x and y are the values within each bias entry. E_B can be also represented as three-element ordered pair, depending upon the location of neuron within the network hierarchical structure. Linear search mechanism is used in searching matched bias entry with the adjacency information obtained for a given input pattern

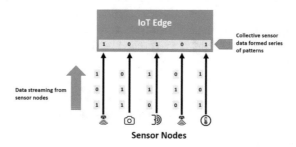

Fig. 2. Distributed pattern recognition approach for event data monitoring and processing.

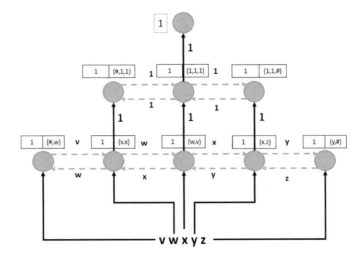

Fig. 3. HGN network composition for storing input pattern of length 5, with string value "v, w, x, y, z".

within each neuron. Each entry within the bias array is unique. Therefore, the following equation is used to estimate the expected number of comparisons for each input entry C_i, given a bias array of size N and r_i number of occurrences for each entry:

$$C_i = \frac{N+1}{r_i+1}; i \in N \tag{3}$$

Within HGN implementation, each neuron performs a forward propagation of index values obtained from the matching process as shown in Fig. 3, from the base layer neurons towards the top neuron within the network composition. The top layer neuron will determine the final index produced by the network. The procedures within the implemented scheme follows a one-pass cycle for each input pattern, without any iterations involving value alteration as to obtain a recognition output.

HGN distributed recognition approach has been chosen as it enables recognition to be performed in a single-cycle procedure, as well as enabling data from different IoT sensors to be collectively analyzed as data patterns. In the following section, we will present the overall framework for event data processing within the IoT-blockchain network that implements the proposed HGN distributed pattern recognition algorithm.

3   Distributed Pattern Recognition Framework for Event Processing in IoT-Blockchain Network

3.1   System Model

As shown in Fig. 4, the proposed IoT-blockchain network comprised of two type of nodes: sensor node and IoT edge node. The network configuration used is based upon the assumption that all the nodes are able to communicate with each other with low latency in a full-mesh network structure. The sensor node acts as the event data collector with limited computational resources. The data update process happens when these sensor nodes broadcast the event data to the IoT edge nodes. It is also assumed that the node is capable of broadcasting its identity to IoT edge nodes using a multicast transmission. For sensor nodes, any communications received from other sensor nodes will be discarded.

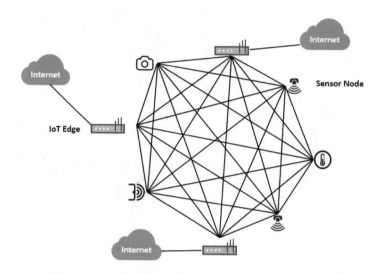

Fig. 4. Illustration of the considered IoT-blockchain system.

In the proposed design, a sensory node S will first establish its identity, and broadcast it to the IoT edge nodes, E. For a given specific time interval T, the sensor nodes will transmit a sensor reading value. This will then be processed by the IoT edge node as data pattern for T. The data patterns will then be analyzed using our proposed HGN distributed recognition algorithm on each IoT edge node. Output of the analysis is in the form of combined pattern index, event status, and event pattern. The event status consists of information whether the event is classified as new or recalled. Only new event will be stored in the blockchain data block, as it contains unique event.

Figure 5 illustrates a scenario where a new event is detected within the IoT-blockchain network with five sensor nodes and three IoT edge nodes. We use binary values to represent triggering event and normal (non-event) input. Given that the event pattern retrieved has not been recorded in the IoT edge node's bias array, the event is classified as new event and will be included in the data block.

Figure 6 shows a scenario where recalled event is being detected. In this situation, the event will not be included in the data block, as it matches one of the bias entries in the array.

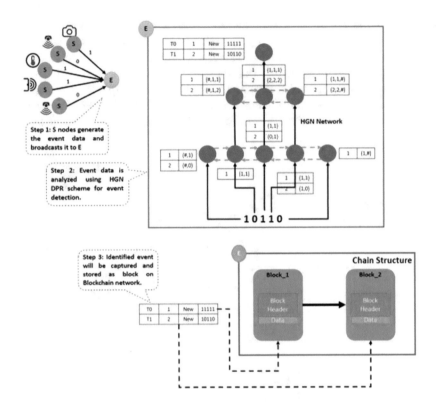

Fig. 5. Illustration of the event processing using HGN distributed recognition scheme on IoT edge node for newly detected event.

4   Simulation and Results

4.1   Simulation Setup

To validate our proposed distributed event processing scheme, we derived a simulation procedure involving event data analysis using the HGN recognition algorithm. In addition, we also perform a blockchain simulation using an online blockchain simulation tool [1]. The simulation involves analyzing the forest fire data, containing 517 records as reported by Cortez and Morais [5]. Each event pattern detected will be stored as a data block on blockchain network. Each data block is verified using the SHA-256 secure hash algorithm.

We extracted 5 important features that contribute towards possibility of fire ignition, namely the ISI index, temperature, relative humidity (RH), wind and rain. For the purpose of simulation, we transform the dataset into a binary format using an interactive binning method. We specified the range of values for each bin according to the potential event (binary value 1) and non-event (binary value 0) as shown in the Table 1.

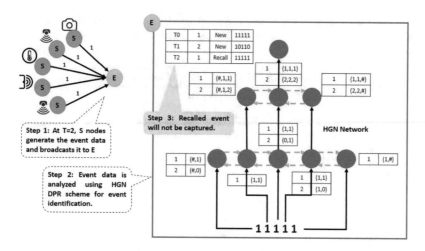

Fig. 6. Illustration of the event processing using HGN distributed recognition scheme on IoT edge node for recalled event.

Table 1. Range of values used to identify each parameter as event.

ISI	Temperature	Relative Humidity (RH)	Wind	Rain
$x \geq 42.075$;	$x \geq 23.325$;	$x \geq 63.75$;	$x \geq 6.75$;	$x \geq 4.8$;
$x \leq 14.025$	$x \leq 7.775$	$x \leq 21.25$	$x \leq 2.25$	$x \leq 1.6$

In implementing this simulation, we assume that the dataset is continuous and taken in a timely manner, starting from time T_0 till T_{516}, respectively.

4.2 Results and Discussion

In this section, we present our preliminary results on event pattern analysis using the Hierarchical Graph Neuron (HGN) approach. The proposed scheme was able to minimize the number of data block generations by 97% of the overall possible block generation on 517 data records retrieved from the forest fire dataset. This was achieved by detecting unique event patterns from the dataset. Table 2 below shows a value comparison between the proposed IoT-blockchain with HGN implementation scheme, against a default event processing scheme.

The proposed HGN algorithm for recognizing patterns on the event dataset is capable of detecting initial 19 event patterns as shown in Fig. 7. However this event patterns are not completely unique, as it contains samples of misclassified events. Note that some duplication exists as pattern with index 7 is reclassified as index 14. In order to eliminate such misclassification, we extend our analysis by examining the confusion matrix of the event classification obtained through our HGN distributed recognition approach, as shown in Fig. 8.

Table 2. A comparison on the number of blocks generated on blockchain network using the proposed IoT-blockchain with HGN implementation and default IoT-blokchain network.

Implementation scheme	No. of blocks generated
IoT-Blockchain with HGN	19
Default IoT-Blockchain	517

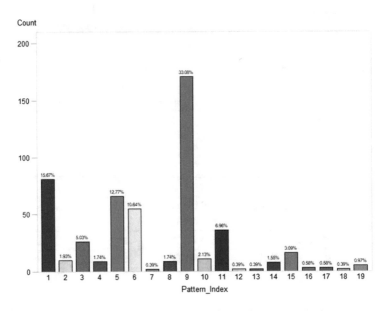

Fig. 7. Initial event pattern distribution retrieved using HGN network on forest fire dataset.

n = 517	Predicted: NO	Predicted: YES
Actual: NO	TN: 0	FP: 0
Actual: YES	FN: 40	TP: 477

Performance Parameter	Value
Accuracy	0.923
Error Rate	0.077
Sensitivity	1

Fig. 8. Confusion Matrix for the event pattern classification using HGN approach.

The actual classification results after the analysis has been carried out indicates that the number of unique event patterns derived from the recognition process is reduced to 15, as shown in Fig. 9. Such reduction is derived as a result of similar patterns being misclassified as patterns with existing recalled indices. For example, patterns with index 7 are 14 are similar, as well as patterns with index 4 and 15.

The simulation work shows significant reduction in the possible number of data blocks generation using our proposed event detection and processing mechanism. Furthermore, the error rate obtained from the event classification is con-

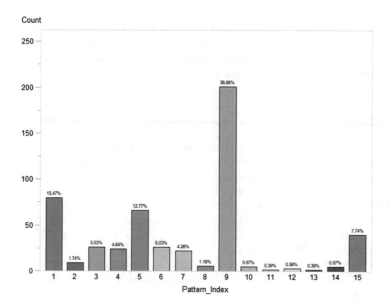

Fig. 9. Final event pattern distribution retrieved using HGN network on forest fire dataset after extended analysis.

siderably low, as it only exerts value around 0.077. Nevertheless, the proposed HGN recognition scheme might be further improved by minimizing the number of neuron layers involved in pattern processing.

Apart from accuracy, HGN algorithm also capable of scaling up to the size of inputs as well as the number of devices attached to a network. Distributed Hierarchical Graph Neuron (DHGN) allows HGN recognition procedure to be performed on multiple-connected nodes as reported by Muhamad Amin and Khan [2]. The interconnectivity between multiple HGN networks allow recognition of events to be conducted at a broader scale, involving heterogeneous IoT networks.

5   Conclusions and Future Work

In this paper, we present our initial implementation of IoT-blockchain network with distributed pattern recognition implementation for event data processing. The proposed scheme implements a HGN recognition approach that minimizes the complex requirements of generating continuous event data blocks on blockchain network. HGN algorithm offers high recognition accuracy for distributed event detection. The detection mechanism considers the state of the entire IoT network, rather than focusing on individual sensor input values. In addition to this, with the distributed ledger mechanism in blockchain technology, such occurrence of events could be stored and verified in a distributed manner.

This paper eventually contributes towards initial implementation of the proposed concept. We intend to further investigate the effectiveness of the proposed scheme, by looking into the scalability of the system when dealing with multi-heterogeneous IoT-blockchain network.

References

1. Blockchain demo. https://blockchaindemo.io/. Accessed 28 Jan 2019
2. Amin, A.H.M., Khan, A.I.: Collaborative-comparison learning for complex event detection using distributed hierarchical graph neuron (DHGN) approach in wireless sensor network. In: Australasian Joint Conference on Artificial Intelligence, pp. 111–120. Springer, Heidelberg (2009)
3. Amin, A.H.M., Khan, A.I., Nasution, B.B.: Internet-Scale Pattern Recognition: New Techniques for Voluminous Data Sets and Data Clouds. Chapman and Hall/CRC, Boca Raton (2012)
4. Buterin, V.: Toward a 12-second block time. Ethereum Blog (2014)
5. Cortez, P., Morais, A.D.J.R.: A data mining approach to predict forest fires using meteorological data (2007)
6. Diedrich, H.: Ethereum: Blockchains, Digital Assets, Smart Contracts, Decentralized Autonomous Organizations. Wildfire Publishing, Sydney (2016)
7. Jan, M.A., Nanda, P., He, X., Liu, R.P.: A sybil attack detection scheme for a forest wildfire monitoring application. Future Gen. Comput. Syst. **80**, 613–626 (2018)
8. Khan, A.I., Mihailescu, P.: Parallel pattern recognition computations within a wireless sensor network. In: Proceedings of the 17th International Conference on Pattern Recognition, ICPR 2004, vol. 1, pp. 777–780, August 2004
9. Kleyko, D., Osipov, E.: On bidirectional transitions between localist and distributed representations: the case of common substrings search using vector symbolic architecture. Procedia Comput. Sci. **41**, 104–113 (2014)
10. Nakamoto, S.: Bitcoin: a peer-to-peer electronic cash system (2008)
11. Nasution, B.B., Khan, A.I.: A hierarchical graph neuron scheme for real-time pattern recognition. IEEE Trans. Neural Netw. **19**(2), 212–229 (2008)
12. Wood, G.: Ethereum: a secure decentralised generalised transaction ledger. Ethereum Proj. Yellow Pap. **151**, 1–32 (2014)

On Value Preservation with Distributed Ledger Technologies, Intelligent Agents, and Digital Preservation

Josep Lluis de la Rosa[1,2(✉)]

[1] TECNIO Centre EASY – Vicorob Institute,
University of Girona, Girona, Spain
peplluis@eia.udg.edu
[2] ETH Zurich, Zurich, Switzerland

Abstract. This is prospect research on digital preservation at the crossroads of blockchain and artificial intelligence with a double aim: first, fulfil a vision of self-preserving digital objects and contribute with a dare step towards the long-term digital preservation, and second, open an area of research that is the preservation of value once we are witnessing the dawn of the internet of value enabled by the advent of blockchain and other distributed ledger technologies .

Keywords: Agents · Digital preservation · Blockchain · DLT ·
Virtual currencies

1 Introduction

For keeping Digital Objects (DOs) genuine and usable for a long time, there is a new need of **digital preservation** (DP) for the prevention or curation of their obsolescence regardless of how they are stored (whether in cloud, local, or distributed and accessible through ledgers) with a series of operations: migration of formats [20] curation whenever errors occur, metadata to understand their context, errors in the early DP strategy, and so on and forth. Unfortunately, these operations derive in costs in maintaining the DP services that are growing in a way to risk turning up unaffordable. A lack of DP resources causes losses in DO usability and integrity and, certainly a loss of value. My proposition was to redistribute the costs of DP so that everyone contributes to the DP of whatever DO of personal, business, administrative, cultural, historical, scientific, educational, artistic, economic, or any other reasons for **value**.

DP strategies so far have been centralised: national archives and libraries play a central role, a central view, where open source solutions with archivists, professionals or amateurs are dominating the DP landscape. An example of a centralised system with crowd contributions was the FP7 DURAFILE project (http://www.durafile.eu - *Innovative Digital Preservation using Social Search in Agent Environments*), where I introduced social search and agent technologies to develop collaboratively preservation plans for DOs [23, 24]. However, **nearly no attempts to fully decentralise and distribute the DP and its costs have been developed** and none of them has been successful yet. A first approach was [13] and years later our we worked out our own

J. Prieto et al. (Eds.): BLOCKCHAIN 2019, AISC 1010, pp. 145–152, 2020.
https://doi.org/10.1007/978-3-030-23813-1_18

approach in MIDPOINT project (http://easy4.udg.edu/midpoint - *Nuevos enfoques de preservación digital con mejor gestión de costes que garantizan su sostenibilidad*), with the birth of cost-aware digital object (CADO). Other examples in the literature were *buckets* to make archives smarter as [13] proposed or *agents* that become aware of obsolescent formats and seek suggestions from other agents [14]. Those years, we settled the foundations of an Object Centric View (OCV) of the DP versus the prevailing central view that intelligent agents fit well as a solution.

Meanwhile, the field of **cloud storage has been fast moving towards distributing the storage of digital content by means of DLTs**: Siacoin [22], Filecoin.io [15], and the protocols IPFS [1] or Akasha.world are examples of fully distributed, peer to peer storage and provide benefits in robustness and in cost reduction. As DP is built-up on top of *storage*, a distributed cloud storage will help addressing several of the said DP issues like usability, integrity, and cost. These early developments of cloud storage with costs reduction and enhanced scalability opened my eyes: there is chance to apply DLT to DP to save similar magnitude of costs reduction. Moreover, it is worth exploring how our OCV might also adopt the DLT to better deal with *value* **when DO are indeed digital assets**.

Thus it is worth revisiting DP with my **OCV with a new background on DLT** to fulfil my vision of DP developed by intelligent, autonomous agents as social (peer to peer) duty [12] living on a ledger, aiming at reducing costs by several orders of magnitude [15] while keeping intact usability and integrity of DOs and their value: The OCV of DP by the adoption of Intelligent Agents paradigm adapted to the DLTs is a paradigmatic view of distributing DP that we expect will impact not only in scalable cost for obsolescence management but in **diminishing the potential value losses of digital assets after being exchanged many times**.

2   Initial Hypothesis and General Objectives Pursued

My first hypothesis is that the OCV of DP developed as smart contracts under **the intelligent agent paradigm can be empowered with DLT to deliver the less costly DP solutions**, aiming at a linear growth of their costs through time instead of an exponential one.

My second hypothesis uses an analogy of the DO obsolescence through time and DO value depletion through a succession of change of hands. In DP, while time is passing, DOs suffer from a sequence of migrations or technology innovations. This degradation process of the DO integrity and other properties happens in parallel to the DO value loss after a sequence of ownerships and licenses: it will be object of Value Preservation (VP). Thus, the second hypothesis is that **DLT/AI based OCV of DP will properly preserve the value of digital assets**.

It is an unchartered territory: *value preservation* (VP). We explore how value is preserved in Intellectual Property (IP) preservation, notably in Open Innovation (OI) and industrial secrets contained in 3D objects model designs (3DDO). Then, we *revisit* the Cost Aware Digital Object (CADO) research and develop the proper intelligence to deal with Long Term Digital Preservation (LTDP) by distributing the preservation logic, efforts, and cost, which means bringing together the benefits of DP together with those of intellectual property preservation (IPP), and develop them on DLT.

On one hand, I seek for DOs that autonomously connect to preservation services. There are two topics that need to be researched in this area:

First topic, there is need to find out whether preservation is provided as a service (e.g. by currently existing DP solutions providers) that can be called remotely through oracles or as smart contracts (SC). In a first case, further R + D is necessary for preservation services being called through oracles, DP solution-providers that offer closed solutions so that third parties call to their services remotely. In a second case, one step further is that preservation services would be implemented as SC as well. This would require an important change in DP solution providers, since anyone able to provide preservation services would be able to create a SC and each node of the DLT would, thus, execute the DP services when the SC are called.

Second topic, it is a key research topic to tackle how to enable that a SC works autonomously. Current DLTs and SC solutions do not enable SC to self-run themselves periodically or after a specific amount of time and they need to rely on external services (either human or software agents) that take charge of calling the SC when needed. Those external calling entities are also responsible for the provision of the funds to run the SC. There is a need to change the paradigm to enable the development of fully autonomous DO relying on SCs. To do so, we bet for creating a fork of one of the open-source existing SC platforms like Ethereum and extending their implementation to support such a kind of operation.

On the other hand, there is further research need in how to improve trust in collaborative environments though IP protection of the DOs that contain an asset to be shared. Here the key aspects to be tackled include ownership, which is solved in the state of the art, but also ownership transfer and inheritance, which need to be further researched. Any DO can be subject to transactions according to its value that might affect the value itself, either monetary losses or misuse losses because of the death of the owner, for example. Any DO needs to include mechanisms to foresee that it has been sold or its rights have been transferred to someone else, because the owner decided so or because a mechanism as SC has initiated the transfer after the owner's death.

Other IP related issues include licensing, access control, and reporting. Here the main research challenge relies on autonomously transferring any resources collected by the DO to the owner's balance. Additional research challenges involve the distribution of encryption keys to access content, which cannot be encoded directly into SC, since SC are publicly accessible. Thus, as already mentioned, some mechanisms are needed to enable a fully autonomous operation of the SC in combination with the presence of specific trusted software components in the device of the licensee.

Finally, we foresee the deployment of a token, the Preserva (PRE). *Preservas* are supplied for the tasks of IP preservation (IPP), curation, ingest, checking acknowledgements, mining, and validation of their IP transfer, management, and more. They are created as the friction (fees) of the IPP services over DOs that are registered/entered on the ledger. PRE are paid as the fee for the uploads, transfers, and more, of the DOs, so that they are next allocated to the acknowledged sources as if they were DO (i.e., they were registered, i.e. uploaded). PRE are also paid for checking license, terms of agreement, implemented as SC and the validation of ordinary transactions, or for keeping the DO in the ledger activities. The PRE are created (supplied) for the fees of IP and DO preservation and/or storage DOs despite of any virtual currency might be provisioned for budgeting the costs of DO IPP. That is why the PRE is a utility token, as it can be used for fees or for the IPP.

3 Agents for Managing Digital Preservation, Its Cost, and the Value of Digital Assets

Said that, let's talk further of the agent approach for costs and value. **First**, reducing the magnitude of LTDP has immediate reduction of cost yet the cost might still be exponentially growing unless we adopt the OCV of DP: We have argued in Sect. 2 that the reduction is achievable by converting the DOs into **smart DOs, intelligent agents that selfpreserve under a budget in virtual currency for DP and VP**. That's it, our works in 2014–15, [16–18] and [19] showed that the smart DO are able to manage their DP costs after we introduced self-preservation skills and goals in their intelligence: we **reduced the exponential growth of cost towards into a flatter one**. This is key for the LTDP. We called them CADO (Cost Aware Digital Object) or SPDO (Self-Preserving Digital Objects). In all, CADO was a bottom-up approach of cost management in DP using e-auctions, contrary to the prevailing top-down approach of the state of the art.

First point, algorithms to be researched and re-implemented under SCs as intelligent agents linked to DO are the electronic auctions (e-auctions) which are still a novel approach to managing costs in DP. It is such a kind of solution that takes advantage of the said OCV developed by means of CADO whereby the DOs manage their self-preservation by maximizing the chances of avoiding obsolescence but at a minimum cost. To accomplish this, we assign a budget that the DOs manage to secure the supply of preservation services at a given cost. Several strategies apply, such as maximal preservation service at all costs or burn low even if the preservation outcome is not perfect. We explore optimizing the budget of SPDO through micro-negotiations of DOs and services, expecting accurate balance of costs and quality of preservation. Specifically, in negotiation, we explored price-based algorithms, like the e-auctions with combinatorial and multi-unit auctions [15]. We compared the expected lifetime of DOs under those 2 e-auction algorithms to decide under what conditions they apply and deliver good results. These results will be revisited in the DLT implementations. We defined then a virtual currency called PRESERVA (₽/PRE) to extemporize and universalize the price assignment of the LTDP services' costs so that we use the same prices now and in the future regardless of the monetary paradigm, as well as provide a budget for CADO in a transparent, easy, and general way. **1₽ is the cost of preserving a DO for 100 years**. In our experiments, we used the m₽ (miliPRE) as the DLT token. Thus, prior to a full DLT implementation, simulation experiments delivered interesting results like the following (baseline in blue, multiunit in red, and combinatorial in green for the y-axis, and the x-axis are the number of technological changes that are multiple of 5 years each, thus 20 changes means 100 years). Next experiments to confirm theses result will be done in emulation first in a testnet and then in the open network with a DLT (Fig. 1):

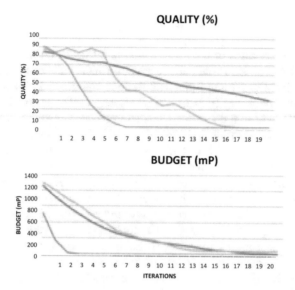

Fig. 1. Quality in % and budget in mPRE of DOs through time, of three e-auctions algorithms

Second point, with the advent of DLTs, the CADO can become a reality as long as we achieve DLTs with **SCs** to support smart and autonomous DOs. I plan to revisit the CADO state of the art with the DLT approach and go far beyond the self-preserving properties through their cost management and introduce other aspects into the study where preservation is also of prominent importance such as the **preservation of value VP**, which is today ill-defined.

We define *value preservation* as "dealing with (explicit) ownership and rights management along with digital preservation of data which, regardless of formats and platforms, is being transferred and processed among entities across public and private bodies". **The preservation of value** means keeping **integrity, genuinity, and usability** of the value related to the DO, for a long term, **across exchanges**. Data, the DOs, turn into digital assets which value, whatever will be, is preserved.

Value is then **an agreement of one or several parties over the digital assets they are using by licensing**. It could be of personal, intellectual, historical, artistic, societal, industrial, or economic dimension of value. I'm talking about handling the VP and not the *value creation*, thus we will just take a theoretical baseline value that will be kept as strong as initially was along the chain of licensing.

Third and last point, being said that one of the dimensions for granting value to DOs is Intellectual Property (IP) and acknowledging there is a growing need to preserve IP [6, 7] as means to preserve the value of DOs, for which ownership (including its transfer and inheritance), licensing, access control, and reporting are relevant aspects that need to be handled. Otherwise, usable and well-preserved DOs from the DP perspective that are bundled to wrong or unauthorised owners **derives into a complete loss of value**, whether for which owners are no longer licensed, accessed or traced.

4 Applicability of the Research

I plan to develop use cases of the *preservation of IP value*, on the domain of online platforms for OI and on industrial trade secrets over industrial designs contained in 3D digital objects (3DDO). **OI and industrial designs contained in 3D objects are two application domains** where IP is of special relevance: The leaks of IP by inappropriately using 3D industrial design is estimated to be causing 200 billion dollar losses only in the US manufacturing industry in favour of Asian countries in counterfeit goods, pirate software, and theft of trade secrets. The fact is the digitization of know-how and online collaboration is making IP leaks much easier than ever. In this line, the lack of effective IP rights management has been identified as one of the main barriers for collaboration to be effective. In the survey [2], the factors acting as hinderers to OI mentioned that beyond the main concern relative to resources –an issue inherent to their SME structural configuration, and their difficulty in finding trustable/reputable partners, **IP issues represent a major concern** [5]. Being OI recently redefined by [4] as the "use of purposive inflows and outflows of knowledge to accelerate internal innovation", it assumes that firms can and should use external ideas as well as internal ideas, and internal and external paths to market, as they seek to improve their performance; this notion of OI is largely based on **transferring knowledge**, expertise, and resources from one company or research institution to another [21], with the risks of **losing value** because of an authorised or inappropriate usages of the digital asset. This is our approach to the preservation of value.

Online OI platforms are territories of huge big data effect, gamification approaches, AI in the automation of ideation and project support services, as well as rewarding schemes with reputation and virtual money, or novel IP management, which need to be researched, developed, and tested in the online OI [3]. I focus on the application cases on **IP value preservation (IPP)** that might play an important role in the development of OI platforms. For the IPP, as I stated in [8] DLTs are expected to open up a range of possibilities as well as challenges. Some of these are already underway, while others may take considerable time due to e.g. regulatory constraints.

There are several IP related applications where DLTs can play an important role [7] that need to be mentioned: *Timestamping, proof of existence or notarization* relies on the use of cryptographic hashes of IP assets to provide proofs of existence. As well, with DLTs, a new wave of *decentralized IP registries* and services have arisen. They can range from just covering authorship recognition, to more sophisticated services enabling access to content, licensing and other features. DLTs and timestamping services can be used together to create an auditable trail of content ownership from creation to the transfer of rights and beyond, while *access control* relies on giving and controlling access to some content to predefined users. It might be used in combination with licensing or NDAs. DLTs enable the implementation of access control mechanisms through the combined use of SCs, record keeping solutions and encryption mechanisms. Examples are Po.et, Creativechain.org, Ascribe.io, and IPSeeds.net. *Licensing* determines the rights and conditions under which someone can make use or access some content or work belonging to the copyright owner. Decentralized licensing relies on SCs. Examples are Creativechain.org as well as the said Ascribe.io and Po.et.

Finally, Non-Disclosure Agreements (NDA) are used to establish a trusted environment for a background knowledge exchange each party is willing to communicate to others and each party recognizes and proves it happened by stamping a signature on. Some initiatives relying on DLTs include IPSeeds.net and Bernstein.io. About *record keeping*, since DLTs are not suitable to keep large amounts of data, alternative mechanisms have been conceived to link files from the DLT itself through the use of hashes and addresses: For example, the IPFS is a P2P protocol under DLT. Finally, thanks to DLTs, it is possible to **design cryptocurrencies to reward or promote behaviours** among the community of knowledge users. Currency designers can determine when new currency is issued and to whom. Existing cases include said Po.et, Creativechain.org, and Witcoin.io

Thus, **through DLTs, I can extend CADO to embrace the DO VP**. If it is put into practice, there will be no need to have centralised systems to manage the VP, and the DOs themselves would be able to act as autonomous objects in the search of their LTDP, keeping their value intact through time.

Our research will contribute to a renewed approach of LTDP, that with the focus on IPP we expect to not only reshape the state of the art of DP, DLT, and intelligent agents.

Acknowledgement. Thanks to **Alastria**, and to the projects **AfterDigital Consultants**: Digitalización del Consultor Digital, RTC-2017-6370-7, CIEN **ServiceChain** (Nuevas tecnologías basadas en blockchain para gestión de la identidad, confiabilidad y trazabilidad de las transacciones de bienes y servicios), and the *grup de recerca consolidat* -ref. **2017 SGR 1648.**

References

1. Rabinovici-Cohen, S., Baker, M.G., Cummings, R., Fineberg, S., Marberg, J.: Towards SIRF: self-contained information retention format. In: Proceedings of the 4th Annual International Conference on Systems and Storage, SYSTOR 2011. ACM, NY, USA (2011)
2. Bikfalvi, A., de la Rosa, J.L., van Haelst, S., Gorini, M., Pelizzaro, A., Haugk, S.: Study report to characterize the target groups in relation to the project topics: SMEs and innovation advisors, intern report (2016). http://dugi-doc.udg.edu//handle/10256/13269. Accessed 11 Nov 2018
3. Bogers, M., Zobel, A.-K., Afuah, A., Almirall, A., Brunswicker, S., and 15 more authors: The Open Innovation Research Landscape: Established Perspectives and Emerging Themes Across Different Levels of Analysis, Forthcoming in Industry and Innovation (2017)
4. Chesbrough, H., Bogers, M.: Explicating open innovation: clarifying an emerging paradigm for understanding innovation keywords. In: Chesbrough, H., Vanhaverbeke, W., West, J. (eds.), New Frontiers in Open Innovation, pp. 1–37. Oxford Press (2014)
5. Chesbrough, H., Ghafele, R.: Open innovation and intellectual property: a two sided market perspective. In: Chesbrough, H., Vanhaverbeke, W., West, J. (eds.) New Frontiers in OI, 191–207. Oxford University Press (2014)
6. de la Rosa, J.L., Gibovic, D., Torres, V.: A preliminary work on virtual currencies for OI. In: IV International Conference on Social and Complementary Currencies, Barcelona, 22–24 May 2017 (2017)

7. de la Rosa, J.L., Torres-Padrosa, V., El-Fakdi, A., Gibovic, D., Hornyák, O., Maicher, L., Miralles, F.: A survey of blockchain technologies for open innovation. In: 4th Annual World Open Innovation Conference WOIC 2017, San Francisco, USA, December 2017

8. de la Rosa, J.L., Gibovic, D., Torres-Padrosa, V., Maicher, L., Miralles, F., El-Fakdi, A., Bikfalvi, A.: On intellectual property in online innovation for SME by means of blockchain and smart contracts. In: 3rd World Open Innovation Conference WOIC 2016, Barcelona, December 15 2016

9. de la Rosa, J.L., Stodder, J.: On velocity in several complementary currencies. Int. J. Commun. Curr. Res. **19**(D), 114–127 (2015)

10. de la Rosa, J.L., Olvera, J.A.: La Preservación Digital Como Asunto Social: Motivación Al Archivo Personal. Tabula: revista de archivos de Castilla y León, ISSN 1132-6506, No. 17, pp. 135–156 (2014)

11. de la Rosa, J.L., Olvera, J.A.: First Studies on Self-Preserving Digital Objects, Frontiers in Artificial Intelligence and Applications – AI Research & Development, ISSN 0922-6389, vol. 248, pp. 213–222. IOS Press, Amsterdam (2012)

12. de la Rosa, J.L.: Analysis of Digital Preservation as a social duty, Internal Deliverable Project, PROTAGE, PReservation Organizations using Tools in AGent Environments, Grant no.: 216,746, FP7 ICT-1-4.1 Digital libraries and tech-enhanced learning (2011)

13. Nelson, M.L., Maly, K.: Buckets: smart objects for digital libraries. Commun. ACM **44**(5), 60–62 (2001)

14. Pellegrino, J.: A multi-agent based digital preservation model. arXiv preprint arXiv:1408. 6126 (2014)

15. Protocol Labs: Filecoin: A Decentralized Storage Network (2017).https://filecoin.io/filecoin. pdf. Accessed 8 Dec 2018

16. Olvera, J.A., Carrillo, P.N., de la Rosa, J.L.: Evaluating auction mechanisms for the preservation of cost-aware digital objects under constrained digital preservation budgets. In: Kapidakis, S., et al. (eds.) Research and A. Tech for Digital Libraries. LNCS, vol. 9316 (2015)

17. Olvera, J.A., de la Rosa, J.L.: Time machine: projecting the digital assets onto the future simulation environment. In: Proceedings of the 13th International Conference on Advances in PAAMS, The PAAMS Collection LNCS, PAAMS 2015, Salamanca, Spain, LNCS vol. 9086, pp 175–186 (2015)

18. Olvera, J.A., de la Rosa, J.L.: Addressing long-term digital preservation through computational intelligence. In: Proceedings of the 13th International Conference on PAAMS 2015, Salamanca, Spain Advances in PAAMS, The PAAMS Collection LNCS vol. 9086, pp. 304–307 (2015)

19. Olvera, J.A., de la Rosa, J.L., Carrillo, P.: Combinatorial and multi-unit auctions applied to digital preservation. In: Museros, L., et al. (eds.) Artificial Intelligence Research And Development, pp. 265–268. IOS Press (2014). https://doi.org/10.3233/978-1-61499-452-7-265

20. Benet, J.: IPFS - Content Addressed, Versioned, P2P File System (2014). https://ipfs.io/ipfs/. Accessed 21 Nov 2018

21. Van de Vrande, V., De Jong, J.P.J., Vanhaverbeke, W., De Rochemont, M.: OI in SMEs: trends, motives and management challenges. Technovation **29**(6), 423–437 (2009)

22. Vorick, D., Champine, L.: SIA: Simple decentralized storage (2014). https://www.sia.tech/ sia.pdf. Accessed 08 Dec 2018

23. Trias, A., de la Rosa, J.L.: Question waves: an algorithm that combines answer relevance with speediness in social search. Inf. Sci. **253**, 1–25 (2013)

24. Trias, A., de la Rosa, J.L.: Survey of social search from the perspectives of the village paradigm and online social networks. J. Inf. Sci. **39**(5), 688–707 (2013)

Blockchain Technology: A Review of the Current Challenges of Cryptocurrency

Diego Valdeolmillos[1] (ID), Yeray Mezquita[1(✉)] (ID),
Alfonso González-Briones[1,2(✉)] (ID), Javier Prieto[1(✉)] (ID),
and Juan Manuel Corchado[1,2,3,4(✉)] (ID)

[1] BISITE Research Group, University of Salamanca,
Calle Espejo 2, 37007 Salamanca, Spain
{dval,yeraymm,alfonsogb,javierp,corchado}@usal.es
[2] Air Institute, IoT Digital Innovation Hub (Spain),
37188 Carbajosa de la Sagrada, Spain
[3] Department of Electronics, Information and Communication,
Faculty of Engineering, Osaka Institute of Technology, Osaka 535-8585, Japan
[4] Pusat Komputeran dan Informatik, Universiti Malaysia Kelantan,
Karung Berkunci 36, Pengkaan Chepa, 16100 Kota Bharu, Kelantan, Malaysia

Abstract. As a result of the 2017 boom in the cryptocurrency market, some governments around the world have begun to work in the direction of regularizing and supervising digital currency. People have gained trust in the use of cryptocurrency thanks to the security of the blockchain technology and of their economic ecosystem. This paper reviews the challenges faced by five different cryptocurrencies with the highest market capitalization. Furthermore, we analyze the blockchain technology that underlies them.

Keywords: Blockchain technology · Legal framework · Cryptocurrencies · Review

1 Introduction

The financial crisis of 2008 has shown that banks and centralized financial institutions have breached the trust of the people who deposit their money in them, by lending it while keeping very little in reserve. Bitcoin emerged in 2009 as an answer to those transgressions, the solution it provided was to have a currency without the need of a central authority [1].

The Bitcoin platform comprises a series of cryptographic protocols that totally transform the way in which transactions are made. Thus, this platform has brought the financial system one step closer to a true democratic economy constructed by the community. The Blockchain Technology (BT), on which Bitcoin operates, implements a distributed ledger across the peer-to-peer network of actors that participate in the system. This technology makes possible to validate and verify the transactions carried out within the Bitcoin platform. Thanks to this, a central body that acts as a trusted intermediary is no longer necessary [15]; the nodes of the blockchain network provide users with confidence and security in carrying out transactions.

© Springer Nature Switzerland AG 2020
J. Prieto et al. (Eds.): BLOCKCHAIN 2019, AISC 1010, pp. 153–160, 2020.
https://doi.org/10.1007/978-3-030-23813-1_19

The use of BT is becoming widespread among applications that require the use of digital records and transactions between non-trusted parties. These applications range from financial services to asset traceability and much more. Their incorporation of BT is innovative because it allows non-trusted parties to reach agreements called smart contracts [5].

In most cases, smart contracts make it possible to establish agreements in an autonomous way and without the need of intermediaries [14, 16, 18]. However, in some cases, due to the limited capacity of the blockchain to interact with the real world, the establishment of those contracts requires an external entity called oracle. The oracle is an intermediary between the data inside the blockchain and the data outside of it [17, 20].

BT-based platforms are a type of Distributed Ledger Technology (DLT) system, but not all DLT systems make use of BT. Every cryptocurrency has its own underlying DLT platform, with distinct features. The variations in the DLT platforms may include the algorithm they employ for consensus between the nodes of the network, the rate at which new coins are created, the maximum size of each block of data where transactions are stored, etc.

DLT can be used in a system to improve robustness against traditional cyberattacks. The problem resides in the fact that only a few consensus algorithms are used by the vast majority of cryptocurrencies leaving the door open to the appearance of new specific cyberattacks against this kind of platforms [26]. Despite it, the fact that the networks and protocols that underlies some cryptocurrencies are somehow safe and sound against those attacks, continue attracting new investors and capital to their market.

Considering the boom and the capitalization of the cryptocurrency market that occurred in 2017, some governments around the world have begun to work in the direction of creating measures for the regularization and supervision of digital currency. A prominent case is that of Malta, whose government has adopted a legal framework in 2018 in which systems based on DLTs are granted legal certainty [6].

This paper reviews some of the cryptocurrencies and the challenges they face. It is structured as follows: Sect. 2 provides background information on BT and some of their actual challenges. Section 3 analyzes five cryptocurrencies with the highest market capitalization. Finally, Sect. 4 concludes this review.

2 Background

BT has been created with the intention of replacing the current, centralized financial system. The authors of [13] claim that BT is capable of replacing intermediaries while ensuring the security of platforms. BT offer resistance to traditional cyberattacks, but as this technology gains widespread adoption, they are being developed new attacks specifically for hacking it.

Distributed Denial Of Services (DDOS) attacks are the most common. A kind of DDOS attack is the Malleability attack, produced when an attacker create a copy of a transaction but with another ID, which makes the user spend double for it [26]. This attack occurs when a system that make use of a blockchain, like a bitcoin exchange, have flaws in the implementation of the code that allows the trading of cryptocurrencies.

The eclipse and sybil attacks have similar bases. In both, the attacker gains control of a large number of IP addresses of the network and surrounds the victim with them. In Eclipse attacks, the victim is not allowed to obtain transactions they are interested in, it has been successfully carried out in the Ethereum blockchain by Researchers from the University of Boston. In a Sybil attack, the victim is influenced by the voting power of the attacker nodes and the information they send to it, which makes the victim vulnerable to double spend attacks.

A 51% or majority attack occurred when a single entity owns the majority of the voting power of a network. An attacker who wants to take advantage of this condition can create a fork of the main chain with the transactions it wants to be done. The small cryptocurrencies are at risk because their networks are composed of few nodes.

The more proven a BT-based platform is against the attacks previously mentioned, the more trust the users give in the cryptocurrency it underlies. As a result of that trust its economic ecosystem will grow, which will translate into an increment in the value of the cryptocurrency [13].

Some of the differential aspects of a BT based platform are: the consensus algorithm that the peers of the network use to add new blocks to the blockchain; the way in which the network is governed; and its capability to execute code that does or does not allow to deploy Turing-complete Smart Contracts in the blockchain.

Relating to the consensus algorithm, there is an increasing number of them and their own variations. The most widespread algorithms are the most proved, that's why they, or a their variations, are shared by the vast majority of the cryptocurrencies [2].

In the Proof-of-Work (PoW) algorithm, to add a new block to the blockchain a cryptographic problem must be solved. The computational cost and the difficulty of solving the problem, the energy spent on searching for its solution (work) and the simplicity of verifying it, are enough reasons to encourage the nodes that wants to add new blocks (miners) not to cheat by adding illegal transactions.

Proof-of-Stake (PoS) is a consensus algorithm, in which miners take turns at adding new blocks. The probability of a miner to receive the turn to add a block depends on the amount of coins deposited as escrow (Stake). This algorithm assumes that a node is going to be honest in order to avoid losing the escrow.

The Practical Byzantine Fault Tolerance (PBFT) the process of adding a new block is called a round. In each round a node is selected to propose a new block, the block needs to receive 2/3 of the votes of all the nodes in the network in order to be valid.

Currently, every consensus algorithm has its own risks and vulnerabilities. For example, PoW wastes a massive amount of energy to produce new blocks. It is very limited in terms of scalability and its mining pools are centralized [23]. In the case of the PoS algorithm, its Nothing at a Stake theory causes to occur forks more frequently in the blockchain than with other consensus algorithms [3]. In the case of PBFT the main risk is that it is a permissioned protocol and not a truly decentralized one [4]. In Table 1 it has been done a comparative between the mentioned algorithms and their three most important characteristics: scalability, consistency and decentralization.

Table 1. Comparison of the consensus algorithms.

Consensus algorithm	Scalable	Consistent	Decentralized
PoW	No	Yes	Yes
PoS	Yes	No	Yes
PBFT	Yes	Yes	No

When there is no consensus in the governance of a blockchain among all the nodes of the network, there is a risk that a hard fork will be produced, splitting the community of the platform. An example of this is when the DAO hack of Ethereum occurred, creating two different blockchains: one called Ethereum classic while the other just Ethereum [7]. This was in effect, a breach of Ethereum's immutability and it left a sizeable minority of the community bitterly dissatisfied.

The possibility of executing Smart Contracts within a blockchain system supports the development of decentralized Applications (dApps) [19, 25]. This feature, if available on a platform, can help grow the economical ecosystem of a cryptocurrency by adding use cases in which the cryptocurrency is used in the form of dApps. This helps attract more users and investors who are eager to use the cryptocurrencies.

The possibility of deploying Turing-complete Smart Contracts appeared with Ethereum [8]. This kind of blockchains are called second generation blockchains and experience the same scalability problems as the first generation blockchains, platforms that do not allow the development of Turing-complete smart contracts. However, some of these BT-based platforms support the development of simple non-Turing-complete smart contracts, like in the case of Bitcoin [21].

The third generation of blockchains has come about to increase the number of users and uses of BT-based systems. These blockchain platforms aim to provide solutions to the scalability problems of previous generation blockchains and allow for the mass adoption of dApps in the daily life of people. Examples of third generation blockchains are Tron, Cardano and EOS.

Right now, investors are speculating with the value of the coins in order to increase their income by exchanging cryptocurrencies of different types. That's why, the more people store their wealth in a cryptocurrency, the more difficult to let single fortunes control the fluctuations of its value [24].

3   Cryptocurrencies and Technologies

Part of the value of cryptocurrencies lies in the characteristics of the BTs on which they are based and in the economic ecosystem that supports them. For this reason, this section describes the distinctive characteristics of five currencies with the highest market capitalization, listed by CoinMarketCap [9].

Bitcoin (BTC). Bitcoin was the first use case of the BT. It has gained trust among cryptocurrency users as it has proven, indisputably, the security of its technology. This explains why it is the most used cryptocurrency in the world. However, it has some downsides too; its consensus algorithm is PoW, meaning that it wastes a lot of energy when adding new blocks to its blockchain and it is not scalable.

Ether (ETH). The Ether cryptocurrency is supported by the Ethereum blockchain. It was the first use case in which smart contracts were deployed and the most used Smart Contract platform. The value of this cryptocurrency comes from the fact that its underlying BT offers the possibility of developing dApps that will help the Ether economic ecosystem grow [8]. The problem with Ethereum is that it uses the same kind of consensus algorithm as Bitcoin, making the dApps impossible to escalate in terms of number of users and mainstream adoption by the people. Also, its high fees are a big problem in the way to deploy a dApp capable of handle micro-payments.

Ripple (XRP). This cryptocurrency is housed in the Ripple blockchain. It is scalable, being able to handle up to 1500 transactions per second, which means it has the potential to replace international payment systems like VISA and costs a fraction less. Ripple makes use of the Ripple Protocol Consensus Algorithm (RPCA). It is a variant of the PBFT which makes use of collectively trusted subnetworks of a larger network of validators [10]. The downside of this Consensus algorithm is that it is more centralized than its competitors. This is not a real problem form this platform, because the XRP cryptocurrency was designed a method of helping banks facilitating cross-border money transfers operations between them.

EOS. The EOS cryptocurrency based BT is one of the third generation blockchains. It has a scalable and feeless blockchain while it makes possible to deploy dApps within it. EOS's consensus algorithm is called delegated Proof of Stake (dPoS), a variation of the PoS in which the nodes of the network vote for their representatives according to their stake of funds, which will be the ones that add blocks to the blockchain. This algorithm doesn't have the Nothing at a Stake problem because, under normal conditions, the block producers cooperate to produce blocks rather than compete, making it impossible for any forks to occur [11].

Tether (USDT). Tether is one of the so-called stable coins. It is called that way because it is a non-volatile and stable cryptocurrency used as a dollar substitute in digital environments. To maintain a one-to-one reserve ratio between this cryptocurrency token and its associated real-world asset, the fiat currency, the USD reserves of Tether Limited, should be equal to or greater than the number of USDT in circulation. Tether is backed up by the Bitcoin blockchain via the Omni Layer protocol. This protocol is a software layer that enables next-generation features in the Bitcoin blockchain [12]. It provides off –chain scalability without the need to change the underlying BT.

To do a comparative analysis of the described currencies, we have created a table which includes a selected set of features of the technologies that underlie them, see Table 1. Distinctive features of the technology have been chosen, such as the consensus algorithm of the BTs; the scalability provided by the BT; and whether it is possible or not to deploy smart contracts within its BT. To show the activity of the ecosystem built around those cryptocurrencies, other features have been obtained from Coin Metrics, like the number of completed Transactions Per Second (TPS) and addresses that were active on January 28th, 2019, [22].

Table 2. Comparison of the features of the described cryptocurrencies.

Cryptocurrency	Consensus algorithm	Scalability	Smart Contracts	Transaction count	Active addresses
Bitcoin	PoW	No, ~7 TPS	Non-Turing-Complete	307.917 K	636.805 K
ETH	PoW	No, ~7 TPS	Turing-Complete	568.564 K	242.449 K
XRP	RPCA	~1500 TPS	Not yet	455.801 K	5.986 K
EOS	dPoS	Yes, ~10 k TPS	Turing-Complete	6.33593 M	91.347 K
Tether	Omni Layer protocol over Bitcoin	Yes, off-chain scalability	Not yet	20.244 K	10.819 K

By looking at Table 2, one may infer that the number of active addresses of a currency is related to its position in the market capitalization ranking. However, XRP is an exception to this rule. The reason for this is that it is not used by individuals, but by large corporations, like banks. Also, the currencies that are most used are the ones whose underlying BT makes use of a PoW consensus algorithm.

It is possible to explain why ETH has more daily transactions than Bitcoin due to the use of dApps. Furthermore, the high transaction count of the EOS platform is explained by the activity of the dApps deployed within it and its dPoS consensus algorithm, which can handle a large number of TPS.

In [22] you can see that the EOS cryptocurrency has an upward trend in the number of active addresses, this means that over time users gain trust in the technology that underlies this currency and as a result the number of active addresses increases.

4   Discussion

The rise of BT has contributed to the appearance of a great number of different cryptocurrencies. Each one of them is supported by a DLT platform, whose technology builds user trust in the cryptocurrency and therefore contributes greatly to its value.

The cryptocurrency capitalization market is growing because more investors are putting their money in it. Some countries, like Malta, have seen the potential of cryptocurrency as the main financial service of the future and have started to create legal regulations in order to offer investors a legal framework that covers cryptocurrency activities.

Although BT is a solution to improve security of the data in a traditional system, it is being targeted by new and specific types of cyberattacks. The most popular consensus algorithms, shared by the majority of the cryptocurrencies, are being modified and updated in order to face the specific vector attacks that are aiming to hack BT based systems. It is shown in the analysis carried out in this paper.

The major problem actual consensus algorithms are facing actually is the impossibility to obtain a platform globally scalable, consistent and fully decentralized. All of them have flaws in some of the listed points. Also, another problem is the governance of the network that underlies a BT platform.

Some of the cryptocurrencies are offering their users the possibility of developing and deploy Turing-complete implemented dApps within their economic ecosystem, bringing more functionality to it apart from the trading of assets. This is a measure that not all the cryptocurrencies allow, for example, from the five analyzed, just two of them allow it.

The two cryptocurrencies with most market capitalization make use of the PoW consensus algorithm. Their number of active accounts and transactions per day indicate that this algorithm offers everything their actual ecosystems need. If the ecosystem would grow, in terms of active accounts or number of transactions realized due to the massive adoption of dApps, there are different consensus algorithms like in the case of EOS or XRP.

The fluctuations in the price of the cryptocurrencies due to speculative movements of capital, difficult their use in the daily life of people. But, because the deflationary nature of the cryptocurrencies, if people store their wealth in cryptocurrencies, they would be more robust against the speculative movements made by single great fortunes.

Acknowledgements. This work was developed as part of "Virtual-Ledger-Tecnologies DLT/Blockchain y Cripto-IOT sobre organizaciones virtuales de agentes ligeros y su aplicación en la eficiencia en el transporte de última milla", ID SA267P18, project cofinanced by Junta Castilla y León, Consejería de Educación, and FEDER funds. The research of Yeray Mezquita is supported by the pre-doctoral fellowship from the University of Salamanca and Banco Santander.

References

1. Nakamoto, S.: Bitcoin: a peer-to-peer electronic cash system. (2008)
2. Zheng, Z., et al.: An overview of blockchain technology: architecture, consensus, and future trends. In: 2017 IEEE International Congress on Big Data (BigData Congress). IEEE (2017)
3. Martinez, J.: Understanding proof of stake: the nothing at stake theory. https://medium.com/coinmonks/understanding-proof-of-stake-the-nothing-at-stake-theory-1f0d71bc027. Accessed 27 Jan 2019
4. Witherspoon, Z.: A hitchhiker's guide to consensus algorithms. https://hackernoon.com/a-hitchhikers-guide-to-consensus-algorithms-d81aae3eb0e3. Accessed 27 Jan 2019
5. Casado-Vara, R., Corchado, J.M.: Blockchain for democratic voting: how blockchain could cast off voter fraud. Orient. J. Comp. Sci. Technol., **11**(1) (2018)
6. Parliamentary Secretariat for Financial Services (Digital Economy and Innovation, Prime Minister Office), Malta a Leader in DLT Regulation. https://www.fff-legal.com/wp-content/uploads/2018/02/FSDEI-DLT-Regulation-Document.pdf. Accessed 21 Jan 2019
7. Mehar, M.I., et al.: Understanding a revolutionary and flawed grand experiment in blockchain: the DAO attack. J. Cases Inf. Technol. (JCIT) **21**(1), 19–32 (2019)

8. Buterin, V.: A next-generation smart contract and decentralized application platform. white paper (2014)
9. CoinMarketCap. https://coinmarketcap.com/. Accessed 16 Feb 2019
10. Schwartz, D., Youngs, N., Britto, A.: The Ripple protocol consensus algorithm. Ripple Labs Inc White Paper 5 (2014)
11. Cox, T.: EOS.IO technical white paper. GitHub repository (2017)
12. Omni Layer. https://www.omnilayer.org. Accessed 30 Jan 2019
13. Hawlitschek, F., Notheisen, B., Teubner, T.: The limits of trust-free systems: a literature review on blockchain technology and trust in the sharing economy. Electron. Commer. Res. Appl. **29**, 50–63 (2018)
14. Casado-Vara, R., González-Briones, A., Prieto, J., Corchado, J.M.: Smart contract for monitoring and control of logistics activities: pharmaceutical utilities case study. In: The 13th International Conference on Soft Computing Models in Industrial and Environmental Applications, pp. 509–517. Springer, Cham, June 2018
15. González-Briones, A., Valdeolmillos, D., Casado-Vara, R., Chamoso, P., Coria, J.A.G., Herrera-Viedma, E., Corchado, J.M.: Garbmas: simulation of the application of gamification techniques to increase the amount of recycled waste through a multi-agent system. In: International Symposium on Distributed Computing and Artificial Intelligence, pp. 332–343. Springer, Cham, June 2018
16. Casado-Vara, R., Chamoso, P., De la Prieta, F., Prieto, J., Corchado, J.M.: Non-linear adaptive closed-loop control system for improved efficiency in IoT-blockchain management. Inf. Fusion **49**, 227–239 (2019)
17. González-Briones, A., Castellanos-Garzón, J.A., Mezquita Martín, Y., Prieto, J., Corchado, J.M.: A framework for knowledge discovery from wireless sensor networks in rural environments: a crop irrigation systems case study. Wirel. Commun. Mob. Comput. (2018)
18. Casado-Vara, R., de la Prieta, F., Prieto, J., Corchado, J.M.: Blockchain framework for IoT data quality via edge computing. In: Proceedings of the 1st Workshop on Blockchain-enabled Networked Sensor Systems, pp. 19–24. ACM, November 2018
19. Casado-Vara, R., Prieto, J., De la Prieta, F., Corchado, J.M.: How blockchain improves the supply chain: Case study alimentary supply chain. Proc. Comput. Sci. **134**, 393–398 (2018)
20. Curran, B.: What are Oracles? Smart Contracts, Chainlink & "The Oracle Problem". Accessed 15 Feb 2019
21. Kaiser, I.: A Decentralized Private Marketplace: DRAFT 0.1
22. Coinmetrics. https://coinmetrics.io. Accessed 20 Jan 2019
23. Beikverdi, A., Song, J.S.: Trend of centralization in Bitcoin's distributed network. In: 2015 IEEE/ACIS 16th International Conference on Software Engineering, Artificial Intelligence, Networking and Parallel/Distributed Computing (SNPD). IEEE (2015)
24. Knowledge@Wharton, How the Blockchain Will Impact the Financial Sector, 16 November 2018. http://knowledge.wharton.upenn.edu/article/blockchain-will-impact-financial-sector/. Accessed 23 Jan 2019
25. Casado-Vara, R., Prieto, J., Corchado, J.M.: How blockchain could improve fraud detection in power distribution grid. In: The 13th International Conference on Soft Computing Models in Industrial and Environmental Applications, pp. 67–76. Springer, Cham, June 2018
26. Bryck, A.: Blockchain attack vectors: vulnerabilities of the most secure technology. https://www.apriorit.com/dev-blog/578-blockchain-attack-vectors. Accessed: 26 Mar 2019

A Blockchain- and AI-based Platform for Global Employability

Vid Keršič[✉], Primož Štukelj, Aida Kamišalić[✉], Sašo Karakatić,
and Muhamed Turkanović

Faculty of Electrical Engineering and Computer Science, University of Maribor,
Koroška cesta 46, 2000 Maribor, Slovenia
{vid.kersic,primoz.stukelj}@student.um.si
{aida.kamisalic,saso.karakatic,muhamed.turkanovic}@um.si

Abstract. An adequate project matching and recruitment process are
of high importance for any company and organization. Many factors, such
as rapid growth, sick leaves, or sudden new skill demands, can influence
the rushed hiring decisions, which could lead to choosing an inadequate
person for a project, and, consequently, produce a negative outcome.
Several active platforms exist, that aim to increase employability on a
global scale while offering potential workers with the ideal job position.
In this paper, we present a solution which combines blockchain technology and AI, while taking advantage of their most important and most
relevant features. Blockchain technology is used to ensure the integrity
of the provided information, while the matchmaking process is assured
by AI.

Keywords: Blockchain · Artificial Intelligence · Job matching ·
Recruitment

1 Introduction

It is always a struggle to find the right person for the right job, while the process
of recruiting the right candidate for the right position remains very cost- and
time-consuming. The growing talent scout industry is undeniable proof of that.
Today modern companies rely on talent scout agencies whose core business is to
have an ever-growing pool of people with the right skills and knowledge in order
to meet the clients' demand. However, there is also the challenge of finding the
right person from the pool effectively, since possible candidates can self-generate
profiles with intentionally or unintentionally correct incorrect information based.
Thus, additional effort is needed to determine reliable profiles. In most cases, the
formal education is the basis on which talent agencies and employers make the
shortlist when given a list of candidates. However, the problem resides when the
skills needed are not yet integrated widely and provided within the educational
system, like recently skills, and knowledge on emerging technologies which are
not yet fully covered in official curriculum.

© Springer Nature Switzerland AG 2020
J. Prieto et al. (Eds.): BLOCKCHAIN 2019, AISC 1010, pp. 161–168, 2020.
https://doi.org/10.1007/978-3-030-23813-1_20

Several active platforms aim to increase the employability on a global scale while using new techniques and technologies, while this paper describes an approach tackling the challenge using AI and blockchain technology. The former is used as a richer matchmaker and search tool, while the latter ensures the information integrity and reliability with transparent, immutable and automatic project work management, as well as a payment solution. Although the platform provides a services adequate for various types of the employment sector, it is probably mostly suited for freelancing jobs, which currently lack a mechanism for structured project agreements and payment options.

The structure of the paper is as follows. Related work is summarized in Sect. 2. The main contribution of the paper is presented in Sect. 3, including the detailed proposed solution. Section 4 provides a discussion and final remarks.

2   Related Work

Several solutions for easier employability were proposed based on new and efficient technology. Some of these solution are, similar to ours, based on emerging technologies, such as blockchain, AI and data mining.

A popular online platform for global employability is Freelancer. Although no advanced technology is used in the platform, it is still mentioned due to its popularity (30 million registered users) [1]. Nevertheless, it has some significant downsides, as high fees for freelancers, as well as publishers [2]. False reviews and job listings are another important issue that cannot be controlled. A similar business model with similar drawbacks is found in many online platforms, such as Fiverr, Truelancer and Upwork [3].

A new approach to global employment was introduced by bitJob using blockchain technology. They created a decentralized platform built on the Ethereum blockchain network, which uses its own token named bitJob token (STU) [4]. The platform is made for students, to help them find relevant jobs during their education. Payments for the jobs are made with cryptocurrency, which has much smaller fees than PayPal or credit card payments [5]. Every finished job is examined by one or two audit judges before it is submitted to the client. This approach prevents false reviews and faked finished jobs.

Use of Artificial Intelligence for employability is found in an online platform PandoLogic. They use machine learning to improve and automate the process of sourcing talented individuals. These algorithms scan user profiles from multiple platforms, such as Facebook, LinkedIn, and GlassDoor [6]. Therefore, more profiles can be examined faster and more efficiently.

In the non-freelancing environment, work history is like job reviews. Sarda et al. addressed the problem of falsified work history and past employment [7]. They intended to solve time-consuming verification of an applicant's listed work experience using Ethereum-based public blockchain. They argue that the proposed approach provides cost-effective and real-time work history verification, privacy-preserving and trustworthy data sharing.

Pinna et al. proposed a decentralized system for handling of temporary employment contracts [8]. Temporary workers play an important role in many companies, so hiring them should be an automated and fast procedure. They proposed the D-ES (Decentralized Employment System). Thanks to the D-ES, they are able to implement temporary employment contracts with smart contracts.

An ability aware neural network approach by Qin et al. tackles the huge amount of information in the job market [9]. Their model enhances Person-Job fit for talent recruitment with the help of the Recurrent Neural Network.

The mentioned platforms do not use both blockchain and AI in the job matching and recruitment process, and there is no proposed implementation of the employability system based on these two technologies.

3 Proposed Platform

This section outlines the platform for global employability. It aims to solve the above presented problems of current solutions (e.g. high fees, data integrity).

3.1 Blockchain

Blockchain is used mainly for data integrity, as well as immutable automated business logic based on smart contracts. The solution is implemented as an Ethereum-based blockchain platform, taking advantage of its smart contract properties. The platform is a reliable source of information which presents many benefits, such as automated payments, the creation of own non-fungible tokens, and managing project's success voting. For the purpose of the payment processes, the platform incorporates a fiat-backed stablecoin (e.g. Tether) [10], which are bought with and backed up by conventional fiat currencies (e.g. USD). Each new project created is a smart contract which requires and locks a predefined amount of cryptocurrency tokens in it. Additionally, we introduce another set of tokens, so-called non-fungible tokens, which are used as proof of experience and reliability. Those tokens are assigned to the user who earned them, but cannot be transferred to any other user. After a project is completed, if the project's owner is not satisfied with the results, a set of audit judges are automatically selected to determine if the project was implemented in line with the requirements. The voting is performed with the help of smart contracts.

3.2 Artificial Intelligence

Artificial Intelligence comes into play when coping with an enormous amount of data, and the decision-making process is not feasible with conventional deterministic algorithms [11]. Machine learning is a subset of the AI, and its main advantage is the ability of self-learning. The algorithm can perform a specific task based on patterns while not being programmed explicitly. Learning algorithms are usually divided into three categories: supervised, unsupervised and reinforcement learning. It is very difficult to find appropriate job candidates manually using conventional search engines and on social media sites, such as LinkedIn, due to the vast amount of data. Therefore, AI can be used to determine the best match for a given task. It removes the need for screening applicants and job offers, which can be both labor-intensive and time-consuming tasks. Our platform incorporates supervised learning, which pairs an input to an output based on the learning input-output pairs. This process can be automated and done without human interactions.

3.3   Architecture

The platform itself is a web based solution incorporating classical front end and back end principle, whereby the back end comprises three main components: (1) a database, (2) Blockchain-based smart contracts, and (3) the AI system. Each component serves its own purpose, and is needed for the platform to be fully functional. The high-level architecture of the platform is presented in Fig. 1. A detailed explanation of each component follows.

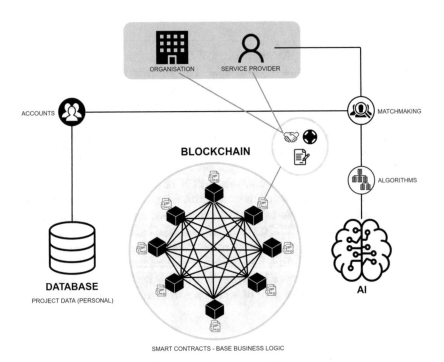

Fig. 1. A high-level depiction of the proposed platform.

The front end of the platform is a **decentralized application (dApp)**. It comprises a simple and easy-to-use website, as well as a communication tool for service providers (e.g. employees) and clients (e.g. companies). Both user groups must register with their personal data to obtain a profile and gain privileges to use the project matching system. The interaction between users occurs entirely over the front end.

Back end consists of three component: (1) **a database**, (2) the implemented business logic based on **smart contracts** deployed on the blockchain network, and (3) the **AI system**. The database itself can be of any type – relational or NoSQL and centralized or distributed. The purpose of the database is to store all the personal data related to the project into the database, thus avoiding storing such data on the blockchain where data is publicly accessible and immutable. Such an architecture is also in compliance with the GDPR rules, where an individual could call for the right to be forgotten. All personal user data can be changed and workers or organizations can update their personal data easily.

Smart contracts are the central part of our system. They control the entire process of creating, storing and executing the system's business logic, and are also used to transfer funds from clients to service providers. Smart contracts contain all the necessary non-personal data for a project to be managed automatically. This mechanism provides a way to save data on the Ethereum blockchain network and, thus, make it immutable. Everyone can also look up the data whenever they want and be sure about data integrity. Before the start of each project, the client transfers cryptocurrency tokens (fiat-backed stablecoins) to a smart contract. This amount is used as a payment to the service providers for a successful project, while some small percentage of the budget is transferred to the platform itself, as a fee for the service. Beside the platform fee, a predefined small percentage of funds is locked for the case of a voting process, whereby the success of the project is decided by trustworthy users, which engage in the voting process due to the incentive of a award payed out from the locked project budget. After the score voting is finished, the smart contracts transfers the project budget according to the predefined rules. Additionally, the service provider would earn non-fungible tokens for a successfully completed project, thus increasing his/her reputation with the second currency in the process, and it is more likely that he/she will be chosen by the AI-based job matching system. The audit judges also earn non-fungible tokens for the same purpose. ERC-721 tokens are also implemented with smart contracts.

The AI system consists of a matching process where the user's projects are matched with potential teams of service providers. If there is more than one match, the client can choose which team will be accepted based on the chance of that team completing the project successfully.

3.4 Pilot Implementation

There are some preconditions for the usage of the proposed platform. Everyone must have an Ethereum account and the MetaMask plugin for communication with the Ethereum blockchain network. With everything set up, users can register on the platform and start using it. After the registration, users must provide personal data for identification purposes, and list types of skills and work history they would like to show under their profile, along with necessary timelines for hiring purposes. On the other hand, individuals or companies seeking an adequate person for a specific project, instead of describing their skill set, they describe the project's idea within the platform.

While the selection of the team is left to the clients, we employ AI to help with the selection of the team. This is done in three stages, (1) providers are assessed according to the requirements of the project, (2) potential teams are constructed in so that all project requirements are met, and (3) all potential teams are evaluated on the possibility of completing the project successfully.

The first stage of the AI is used to evaluate the providers, which is done with the experiences on previous projects on the platform. The blockchain contains information about every experience, the type of work and how many skill tokens were given. The aggregation of all this is used to construct the profile of providers, where skill tokens of the same type of work are summed together. From this, the distance between providers and the project is calculated, which is the coverage between the provider's experience and the required project skills as is used in the calculation of the distance in *k-nearest neighbors* machine learning algorithm.

Next, the selection of providers is made based on their coverage of experiences and the work requirements of the project. From the selection, all possible teams are constructed so that the skill set of all providers cover all of the project requirements, while still taking the provider availability and the limit of skill tokens into account. The team overall skill set is aggregated to form a team vector of experience (the sum of skill tokens), which is accompanied with the information about the size of the team and the duration of the project.

The final stage of the AI is the evaluation of all teams, where the supervised classification technique is used to determine which teams are predicted to finish the project successfully. The decision tree CART classification method is used, which returns the probabilities of predicted classes, which is used in the evaluation of teams. The model is periodically trained on the data for the past time frame, to make the AI system robust to the changing projects and providers on the platform. The input to the classification is the team vector (the aggregate of the skills), and the output is the chance of the successful project. In the current state, the results of the team evaluations are presented to the clients, who choose the team to work on their project, which could be automated in the future.

After the project job matching process is finished, providers are paired with clients, and vice versa. Clients examine the profiles of recommended providers, who have the required skill set, and initiate the bidding process, in which both parties reach a price consensus for the service ordered. Organizations must transfer the stated amount of cryptocurrency tokens into the smart contract immediately upon the agreement. Tokens are locked until the successful completion of the service or the end of the contract. Therefore, both parties are protected. Additionally, all the data related to the service is stored in the smart contract, which brings the features of transparency and immutability by using the blockchain.

Once the provider gives the ordered service within the agreed time frame, the organization must confirm that the provided service is done according to the requirements and finished in the stated time frame. If the service is confirmed, the service provider earns skill tokens, which show the quality of completed services, and increase his/her reputation within the platform. If there are no complaints from the client, the smart contract transfers the cryptocurrency tokens automatically to the provider's wallet as a payment for his/her service. However, if the organization is not satisfied with the delivered service, the voting process starts, where users of the platform are selected as voters, who are independent of the service. The AI system chooses them. The incentive for the voters to actually perform their task fair and fast is the fact that they get a percentage of the project budget. By providing this service, the users also gain skill tokens, which is again used by the system. The number of chosen users should be an odd number to prevent a tie. Voters then decide if the worker has provided the ordered service in terms of the agreement. If the majority of users vote positively, the provider receives skills and cryptocurrency tokens, otherwise, the client gets those tokens back from smart contract, and the provider loses some of his/her reputation within the platform. Some small amount of the payment tokens is transfered to the voters and some small percentage of the budget is transferred to the platform, as a fee for the service.

4   Discussion and Conclusions

The proposed platform combines blockchain and AI with the aim of solving the important employability issues. It offers automated job searching and a recruitment process while preserving the integrity of provided data. The system can lower expenses and shorten the time dedicated to a project matching process. It also provides equal employment opportunities for jobseekers as providers all over the world, without any type of discrimination. Additionally, since it is not using a traditional payment method, it is lowering the fees for payments.

Nevertheless, there are also some features of the presented system which could be seen as drawbacks. The automated smart contract payment mechanism, which is vital for the solution, is currently not acceptable by the majority of legislations. However, the technological field is novel and some legislative initiatives on this topics are already ongoing. It should be noted that the proposed system is a voluntary based one, where service seekers and providers both have to agree on the terms the proposed solutions, thus being legit in the case of an agreement. An AI-driven system used as a part of the recruitment process may cause distrust among users [12]. Therefore, it is essential to demonstrate the system's advantages for the end users. The presented implementation focused on machine learning as an AI tool, which belongs to the statistical learning methods. However probabilistic methods for uncertain reasoning could also be used. Bayesian networks are a known representative of the latter methods and could be used as a matching tool. While decision tree is a useful algorithm for our platform, it is not the only possible choice. In the future, we will explore the possibility of using other machine learning algorithms, such as different variants of logistic regression, linear discriminant analysis, random forest... The efficiency of the used AI algorithm can vary; thus, different algorithms will be tested, and, regarding the outcomes, those that bring the highest efficiency will be used.

Acknowledgments. The authors acknowledge the financial support from the Slovenian Research Agency (Research Core Funding No. P2-0057) and the ECTA initiative.

References

1. Hire Freelancers & Find Freelance Jobs Online - Freelancer. https://www.freelancer.com/. Accessed 13 Feb 2019
2. Freelancer Fees and Charges, Earn And Save More!—Freelancer. https://www.freelancer.com/feesandcharges/. Accessed 13 Feb 2019
3. Best Freelance Platforms in 2019—G2 Crowd. https://www.g2crowd.com/categories/freelance-platforms. Accessed 13 Feb 2019
4. BitJob - students job marketplace—blockchain technology—online jobs. https://bitjob.io/. Accessed 13 Feb 2019
5. faq's—bitJob—STU—blockchain technology—digital currency. https://bitjob.io/questions-and-answers/. Accessed 14 Feb 2019
6. PandoLogic - Pioneering Programmatic Recruitment. https://www.pandologic.com/. Accessed 14 Feb 2019
7. Sarda, P., Chowdhury, M.J.M., Colman, A., Kabir, M.A., Han, J.: Blockchain for fraud prevention: a work-history fraud prevention system. In: 2018 17th IEEE International Conference On Trust, Security And Privacy In Computing And Communications/12th IEEE International Conference On Big Data Science And Engineering (TrustCom/BigDataSE), August 2018. https://ieeexplore.ieee.org/document/8456149
8. Pinna, A., Ibba, S.: A blockchain-based decentralized system for proper handling of temporary employment contracts. In: Advances in Intelligent Systems and Computing, vol. 857, September 2016. https://doi.org/10.1007/978-3-030-01177-2_88
9. Qin, C., Zhu, H., Xu, T., Zhu, C., Jiang, L., Chen, E., Xiong, H.: Enhancing person-job fit for talent recruitment: an ability-aware neural network approach. In: 41st International ACM SIGIR Conference on Research and Development in Information Retrieval, SIGIR 2018, pp. 25–34, July 2018. https://dl.acm.org/citation.cfm?id=3210025
10. Seigneur, J., D'Hautefort, H., Ballocchi, G.: Use case of linking a managed basket of fiat currencies to crypto-tokens. In: First Meeting of the ITU Focus Group on Digital Currency including Digital Fiat Currency. Beijing, China., October 2017. https://archive-ouverte.unige.ch/unige:97657
11. Rusell, S., Norvig, P.: Artificial Intelligence: A Modern Approach, 3rd edn. Prentice Hall, New Jeresy (2010)
12. van Esch, P., Black, J.S., Ferolie, J.: Marketing AI recruitment: the next phase in job application and selection. Comput. Hum. Behav. **90**, 215–222 (2019). https://www.sciencedirect.com/science/article/pii/S0747563218304497

Blockchain and Biometrics: A First Look into Opportunities and Challenges

Oscar Delgado-Mohatar$^{(\boxtimes)}$, Julian Fierrez, Ruben Tolosana,
and Ruben Vera-Rodriguez

Escuela Politecnica Superior, Universidad Autonoma de Madrid, Madrid, Spain
{oscar.delgado,julian.fierrez,ruben.tolosana,ruben.vera}@uam.es

Abstract. Blockchain technology has become a thriving topic in the last years, making possible to transform old-fashioned operations to more fast, secured, and cheap approaches. In this study we explore the potential of blockchain for biometrics, analyzing how both technologies can mutually benefit each other. The contribution of this study is twofold: (1) we provide a short overview of both blockchain and biometrics, focusing on the opportunities and challenges that arise when combining them, and (2) we discuss in more detail blockchain for biometric template protection.

Keywords: Blockchain · Biometrics · Security · Privacy · Vulnerability

1 Introduction

Among all current disruptive technologies, both blockchain and biometrics have become a focus of attention in recent years. On the one hand, blockchain technology provides an immutable and decentralized data registry, optionally with the capability of executing distributed secure code. Its origins are linked to Bitcoin cryptocurrency, created in 2009, where it is used for solving an old problem opened since the 80's in the cryptographic community: the design of a distributed algorithm of consensus on economic transactions without the participation or existence of a central authority [21]. However, nothing prevents any other digital data from being stored instead of economic transactions. This aspect opens the doors to many different potential applications such as smart energy and grids [1,19], healthcare [13], and smart devices or digital identity schemes [22], among others.

On the other hand, the aim of biometric technology is to authenticate the identity of subjects through the use of physiological (e.g., face, fingerprint) or behavioral (e.g., voice, handwritten signature) traits [15]. Its advantages over traditional authentication methods (e.g., no need to carry tokens or remember passwords, they are harder to circumvent, and provide at the same time a stronger link between the subject and the action or event) have allowed a wide

© Springer Nature Switzerland AG 2020
J. Prieto et al. (Eds.): BLOCKCHAIN 2019, AISC 1010, pp. 169–177, 2020.
https://doi.org/10.1007/978-3-030-23813-1_21

deployment of biometric systems, including large-scale national and international initiatives [4,6].

Combining blockchain and biometrics could potentially have many advantages. As a first approximation, the blockchain technology could provide biometric systems with some desirable characteristics such as **immutability**, **accountability**, **availability** or **universal access**:

- By definition, a blockchain guarantees the *immutability* of the registers it stores[1], which could be used by a biometric system to build a secure template storage.
- Derived from previous property, a blockchain increases the *accountability* and *auditability* of the stored data, which can be very useful to demonstrate to a third party (e.g., a regulator) that the biometric patterns have not been modified.
- For last, a (public) blockchain also provides complete *availability* and *universal access* for any user.

Table 1. Blockchain/biometrics mutual benefits

Blockchain to biometrics	Immutability
	Accountability
	Availability
	Universal access
Biometrics to blockchain	More secure digital identity models
	New use cases (e.g., *smart devices*)
	Biometric wallets

Additionally, the integration of biometric technology would be very beneficial for blockchains too. Among many other new use cases, biometrics could greatly improve the current distributed digital identity schemes based on blockchain. Another interesting application of biometrics to blockchain is related to *smart devices*. A smart device is any digital or physical asset with access to a blockchain that can perform actions and make decisions based on the information stored there. For example, a car could be fully managed (rented or bought) through a smart contract. However, an adequate identification of the user is not fully solved yet. In this case, an authentication protocol based on biometrics could significantly raise the current security level. Table 1 overviews the mutual benefits of blockchain and biometrics.

The main contributions of this paper can be summarized as follows:

- We provide a short overview of both blockchain and biometrics, focusing on the opportunities and challenges that arise when combining them.

---

[1] Strictly speaking, a blockchain is not a tamper-proof mechanism but tamper-evident.

– We discuss in more detail an architecture for biometric template protection based on blockchain.

Table 2. Characterization of main blockchain platforms

Blockchain type	Public	Consortium	Private
Governance	No centralized management	Multiple organizations	Tipically, single organization
Access control	Permissionless	Permissioned	
Participants	Anonymous	Identified, trusted	
Main platform	Bitcoin, Ethereum	Quorum, Parity	Hyperledger
Consensus algorithm	PoW / PoS	Voting or multi-party consensus algorithms (PoS /PoA)	
Transaction confirmation time	Long (minutes)	Short	Short
Data privacy	No	Optional	Yes
Smart contracts support	Very limited	Yes	Yes
Cryptocurrency	BTC	ETH	-

The remainder of the paper is organized as follows. In Sect. 2 a description of the most relevant features of blockchain for biometric technologies is provided. In Sect. 3, we first analyze the challenges and limitations of the technologies to finally discuss blockchain for biometric template protection. Finally, Sect. 4 draws the final conclusions and points out some lines for future work.

2 Blockchain Basics

2.1 Overview

Essentially, a blockchain is a decentralized public ledger of all data and transactions that have ever been executed in the system [23]. These transactions are recorded in blocks that are created and added to the blockchain in a linear, chronological order (immutable). Each participating node in the network has the task of validating and relaying transactions, and has a copy of the blockchain.

However, since its initial application to Bitcoin cryptocurrency, the original idea of a universal and public blockchain has greatly evolved into new architectures, based on different access control schemes or consensus algorithms.

According to the first criteria, blockchains can be categorized as: (1) public, (2) consortium, and (3) private blockchains (see Table 2). Essentially, public blockchains are permissionless schemes, designed with a built-in economic incentive for allowing anonymous and universal access. Consortium blockchains, on the other hand, are permissioned, partly private and semi-decentralized architectures, specially targeted for scenarios with a small number of participants. Last, private blockchains are specially indicated in applications where users must be fully identified and trusted. This application environment for private blockchain makes more straightforward the incorporation of biometrics compared to public and consortium blockchains. Anyway, in all three types of blockchains (public, consortium, and private) further research and new security architectures are needed to deliver the full potential of the excellent synergies between blockchain and biometrics.

Blockchains can also use different consensus algorithms, some of which allow greater efficiency and faster transactions completion time. Therefore, the most appropriate type of blockchain depends on the specific use case.

2.2   Smart Contracts

The term *smart contract* dates back to 1996, long before the creation of Bitcoin and blockchain, and was first introduced by Nick Szabo [24]. A smart contract is, essentially, a piece of code executed in a secure environment that controls digital assets. Examples of these secure environments include regular servers controlled by "trusted parties", decentralized networks (blockchains), or servers with secure hardware (SGX) [16,18].

Many public blockchains support the execution of smart contracts, but the most reliable, secure, and used is, without doubt, Ethereum [5]. Ethereum could be considered as a distributed computer, with capability to execute programs written in Turing-complete, high-level programming languages. These programs are no more than a collection of pre-defined instructions and data that has been recorded at a specific address of a blockchain.

For biometric purposes, a smart contract running in a blockchain can assure a semantically correct execution. However, the consensus algorithms necessary to provide this security in public blockchains have an associated economic cost, which will be analyzed in next section.

3   Blockchain for Biometrics

3.1   Challenges and Limitations

Despite the new opportunities already described in previous sections, the combination of both blockchain and biometric technologies is not straightforward due to the limitations of the current blockchain technology. Among them, it is important to remark: (1) its transaction processing capacity is currently very low (around tens of transactions per second), (2) its actual design implies that all system transactions must be stored, which makes the storage space necessary for its management to grow very quickly, and (3) its robustness against different types of attacks has not been sufficiently studied yet.

We now detail the limitations of blockchain public networks for the deployment and operation of biometric systems.

- **Economic cost of executing smart contracts:** In order to support smart contracts in blockchains (like Ethereum), and to reward the nodes that use their computing capacity to maintain the system, each instruction executed requires the payment of a fee in a cryptocurrency (called gas). Simple instructions (such as a sum) cost 1 gas, while others can cost significantly more (e.g., the calculation of a SHA3 hash costs 30 gas). On the other hand, the storage space is especially expensive (around 20k gas for every 256 bits). Therefore, one of the first research problems would be minimizing the cost of running a biometric system (totally or partially) in a blockchain, and how efficiently smart contracts involving biometrics could be coded.

- **Privacy:** By design, all operations carried out in a public blockchain are known by all the participating nodes. Thus, it is not possible to directly use secret cryptographic keys, as this would reduce the number of potential applications. Regarding privacy in public blockchains, three main layers are considered in general: (1) participants, (2) terms, and (3) data. The first one ensures participants to remain anonymous both inside and outside of the blockchain. This is achieved with cryptographic mechanisms like ring signatures, stealth addresses, mixing, or storage of private data off-chain. Second, privacy of terms keeps the logic of the smart contracts secret, by using range proofs or Pedersen commitments. Last, and the most important for biometrics, the data privacy layer goal is to keep transactions, smart contracts, and other data such as biometric templates, encrypted at all times, both on-chain and off-chain. The cryptographic tools used include zero-knowledge proofs (ZKP) and zk-SNARKS, Pedersen commitments, or off-chain privacy layers like hardware-based trusted execution environments (TEEs). However, the application of these cryptographic tools are still very limited for blockchains. For example, Ethereum just included at the end of 2017 basic verification capabilities for ZKPs. More advanced cryptographic tools have been only developed to target special cases like Aztec [25] or ZK range proofs [17]. In addition, it should be noted that ZKP transactions would be still expensive and computationally intensive ($\sim$ 1,5M gas/verification).

- **Processing capability:** Another important limitation is related to its processing capability. Ethereum, for example, is able to run just around a dozen transactions per second, what it could be not enough for some scenarios. Additionally, there is a minimum confirmation time before considering that the transaction has been properly added to the blockchain. This time can oscillate among different blockchains, from tens of seconds to minutes, reducing its usability for biometric systems.

- **Scalability:** This is one of the main handicaps of the technology from its origins as, theoretically, all nodes of the blockchain network must store all blocks of the blockchain network. Currently, the size of the public blockchains (Bitcoin and Ethereum) is around 200GB, and it is growing very fast. This can be a problem for some application scenarios such as the Internet of Things (IoT).

– **Security:** As novel technology, blockchain security characterization is still a work in progress. Among all possible attacks, it is worth mentioning the attack known as *51% attack* [8]. If an attacker gains more than 50% of the computational capacity of any public or private blockchain, he could reverse or falsify transactions. This attack applies even to blockchain with consensus algorithms not based in proof-of-works schemes, like PoS or PoA, typically used in private or consortium topologies. However, the main security problems suffered to date by blockchains are mainly related to programming errors, e.g., the DAO attack happened in 2016, which put at risk the whole Ethereum ecosystem [2].

3.2　Blockchain for Biometric Template Protection

Biometric systems have for long been known to be vulnerable to certain physical [14] and software attacks [11]. Physical attacks to the biometric sensor can be overcome to some extent with presentation attack detection techniques [20]. On the other hand, an important group of software attacks can be prevented using biometric template protection techniques, but the state-of-the-art there [10,12] is still improvable in many ways [9].

Figure 1 depicts the typical stages of a biometric system (in solid gray), all possible points where a biometric system can be attacked, and a representation of biometric template protection based on blockchain (stripped block). By substituting the traditional template store by a blockchain, the security level of the resulting biometric system is significantly increased. If correctly implemented, attacks number 6 (channel interception) and 7 (templates modification) and are no longer possible.

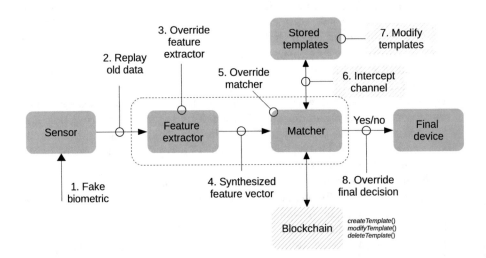

Fig. 1. Main security vulnerabilities of biometric systems and biometric template protection based on blockchain.

This design provides some advantages:

- The modifications to the existing biometric systems are kept to a minimum, so the usual biometric techniques and algorithms (e.g., feature extraction and matching) can be used normally.
- Since the biometric process is performed off-chain, this architecture avoids the scalability problems of public blockchains (except in a massive batch of user registration).
- No need to use complex smart contracts, which facilitates development and reduces execution costs. Smart contracts do not implement biometric "logic", but only the minimum necessary functions to manage the storage of templates (creation, modification, etc.)

However, as stated before, storage space in blockchains is specially expensive compared to computation, in order to discourage its abusive use. As an example, for current Ether price (around 140$ at time of writing, February 2019), a 1KB fingerprint template would cost around 0.00093$ to be stored in Etherum. In any case, blockchains do not usually store data directly, but use distributed storage platforms like IPFS [3].

4   Conclusions and Future Work

Although research on the integration of biometric systems and blockchains is incipient and is taking its first steps, it is undeniable that both technologies have a potential for collaboration and enormous mutual growth.

In this paper we have discussed the main characteristics and limitations of blockchains, especially those that could directly affect the implementation of biometric systems. We have also explored the potential mutual benefits for both technologies, and discussed a first approximation to a combined architecture using blockchain for biometric template protection.

With a view in the future, a key question arises: how many of the biometric processes can be integrated or ported into a blockchain, i.e., done on-chain?. For example, would it be possible to implement a biometric matcher using a smart contract? how? which challenges should be solved to do so?.

Due to the current limitations and characteristics of the blockchain technology, a full integration with biometric processes seems very challenging in the short term. However, there are some promising research areas, e.g., the use of **state channels** [7], which could drastically reduce costs and improve bandwidth, or the development of new zero-knowledge proofs that would allow a user to be authenticated through biometrics without any of the parties having knowledge of the user's identity.

Acknowledgements. Research supported by project BIBECA (RTI2018-101248-B-I00), UAM-CecaBank chair on Biometrics, and UAM-GrantThornton chair on Blockchain. Ruben Tolosana is supported by a FPU Fellowship from Spanish MECD.

References

1. Aggarwal, S., et al.: Energychain: enabling energy trading for smart homes using blockchains in smart grid ecosystem. In: Proceedings of SmartCitiesSecurity (2018)
2. Nicola, A., et al.: A survey of attacks on ethereum smart contracts SoK. In: Proceedings of International Conference on Principles of Security and Trust. Springer (2017)
3. Benet, J.: IPFS - Content Addressed, Versioned, P2P File System, July 2014
4. European Commission: Smart Borders: for an open and secure EU (2013)
5. Dannen, C.: Introducing Ethereum and Solidity: Foundations of Cryptocurrency and Blockchain Programming for Beginners. Apress, Berkeley (2017)
6. Daugman, J.: 600 million citizens of India are now enrolled with biometric ID. SPIE Newsroom (2014)
7. Dziembowski, S., et al.: General state channel networks. In: Proceedings of ACM SIGSAC Conference on Computer and Communications Security, CCS 2018 (2018)
8. Eyal, I., Sirer, E.G.: Majority is not enough: Bitcoin mining is vulnerable. Commun. ACM **61**(7), 95–102 (2018)
9. Fierrez, J., et al.: Multiple classifiers in biometrics. part 2: trends and challenges. Inf. Fusion **44**, 103–112 (2018)
10. Gomez-Barrero, M., et al.: Multi-biometric template protection based on homomorphic encryption. Pattern Recogn. **67**, 149–163 (2017)
11. Gomez-Barrero, M., Galbally, J., Fierrez, J.: Efficient software attack to multimodal biometric systems and its application to face and iris fusion. Pattern Recogn. Lett. **36**, 243–253 (2014)
12. Gomez-Barrero, M., Galbally, J., Morales, A., Fierrez, J.: Privacy-preserving comparison of variable-length data with application to biometric template protection. IEEE Access **5**, 8606–8619 (2017)
13. Gordon, W.J., Catalini, C.: Blockchain technology for healthcare: facilitating the transition to patient-driven interoperability. Comput. Struct. Biotechnol. J. **16**, 224–230 (2018)
14. Hadid, A., Evans, N., Marcel, S., Fierrez, J.: Biometrics systems under spoofing attack: an evaluation methodology and lessons learned. IEEE Sig. Process. Mag. **32**(5), 20–30 (2015)
15. Jain, A.K., et al.: 50 years of biometric research: accomplishments, challenges, and opportunities. Pattern Recogn. Lett. **79**, 80–105 (2016)
16. Karande, V., et al.: SGX-log: securing system logs with SGX. In: Proceedings of ACM Asian Conference on Computer and Communications Security (2017)
17. Koens, T., Ramaekers, C., Van Wijk, C.: Efficient Zero-Knowledge Range Proofs in Ethereum. Technical Report (2018)
18. Küçük, K.A., et al.: Exploring the use of Intel SGX for secure many-party applications. In: Workshop on System Software for Trusted Execution (2016)
19. Magnani, A., et al.: Feather forking as a positive force: incentivising green energy production in a blockchain-based smart grid. In: ACM Workshop on Cryptocurrencies and Blockchains for Distributed Systems (2018)

20. Marcel, S., Nixon, M., Fierrez, J., Evans, N. (eds.): Handbook of Biometric Anti-Spoofing - Presentation Attack Detection, 2nd edn. Springer, Heidelberg (2019). https://doi.org/10.1007/978-3-319-92627-8
21. Nakamoto, S.: Bitcoin: a peer-to-peer electronic cash system (2008)
22. Stokkink, Q., Pouwelse, J.A.: Deployment of a blockchain-based self-sovereign identity. *CoRR*, abs/1806.01926 (2018)
23. Swan, M.: Blockchain: Blueprint for a New Economy. O'Reilly, Sebastopol (2015)
24. Szabo, N.: Smart contracts: building blocks for digital markets (1996)
25. Zachary, J.: Williamson. The AZTEC Protocol. Technical Report (2018)

Author Index

© Springer Nature Switzerland AG 2020
J. Prieto et al. (Eds.): BLOCKCHAIN 2019, AISC 1010, pp. 179–180, 2020.
https://doi.org/10.1007/978-3-030-23813-1

Printed in the United States
By Bookmasters